CHINA

The Rising Aerospace Power

Implications for India

A project under
USI of India, Flying Officer Amandeep Singh Gill
Chair of Excellence

CHINA
The Rising Aerospace Power
Implications for India

Air Marshal Anil Chopra, (Retd), PVSM, AVSM, VM, VSM

Foreword by

Air Chief Marshal RKS Bhadauria, PVSM, AVSM, VM, ADC
Chief of the Air Staff

UNITED SERVICE INSTITUTION OF INDIA
NEW DELHI

PENTAGON PRESS LLP

CHINA—The Rising Aerospace Power: Implications for India
By Air Marshal Anil Chopra, (Retd), PVSM, AVSM, VM, VSM

First Published in 2020

ISBN 978-93-90095-14-8

Published by
PENTAGON PRESS LLP
206, Peacock Lane, Shahpur Jat,
New Delhi-110049
Phones: 011-64706243, 26491568
Telefax: 011-26490600
email: rajan@pentagonpress.in
website: www.pentagonpress.in

Printed at Aegean Offset Printers, Greater Noida, U.P.

Contents

Acknowledgements

The Indian Air Force (IAF) is the primary custodian of air power in India, and it's motto "Touch the Sky with Glory", says it all. I feel very privileged that the Chief of the Air Staff of the Indian Air Force, Air Chief Marshal RKS Bhadauria, PVSM, AVSM, VM, ADC has written the foreword of the book. His appreciation and visionary words mean a lot.

I owe a sense of deep gratitude to the Director, United Services Institute (USI), Maj Gen BK Sharma, AVSM, SM & Bar, and his predecessor, Lt Gen PK Singh PVSM, AVSM (Retd) and the directing staff of USI for their continued guidance, support, and encouragement throughout the writing of this book. I also want to thank all the discussants and participants during presentations for their valuable inputs in channelling my thought process.

I am extremely grateful to the family of the Late Flying Officer Amandeep Singh Gill who have instituted a special chair at USI in memory of the great Air Warrior, under which the research project for the year was entrusted to me.

As I was writing this book, the world went into a terrible pandemic, the corona virus disease (COVID-19). Ironically, it started in China, the country whose aerospace power this book is about. The world had to go into a mass lockdown. This lockdown in a way helped me find time to complete the book. But, unfortunately, many innocent lives were lost worldwide. My salute and respects to those who moved on due to this deadly virus.

My sincere thanks are due to the entire team at Pentagon Press LLP, the publishers of this book, for supporting my raw script to become a book.

Finally, I would like to thank my parents for the great upbringing and seeds of analytical thought, and my wife and children for their encouragement and support when I was busy researching.

Air Marshal Anil Chopra, (Retd),
PVSM, AVSM, VM, VSM

Foreword

The evolution of modern war fighting has established primacy of air power as a primary instrument for nation-states whenever faced with conflict situations. Advancement in aerospace technology has profoundly altered the manner in which modern wars are prosecuted. Use of the third dimension affords significant advantages in terms of responsiveness, flexibility and versatility. Whenever employed in the battle space, air power leaves far reaching effects at the strategic, operational and tactical levels of war. For this reason, air power has become the dominant form of military power projection in the modern world.

China's investment in technological developments and comprehensive military modernisation programs has elevated its combat potential. The Chinese political hierarchy has realised that developing leading aerospace systems is the way ahead. Towards this, the Chinese have aggressively acquired next-gen technologies while concurrently working on concepts of strategic coercion through independent and concentrated use of air power.

In recent times, People's Liberation Army Air Force (PLAAF) has gone through significant structural changes. To begin with, it has started to decouple from the Army centric thinking and influence of the PLA. A significant development in this regard is the presence of two high-ranking PLAAF officers in the Central Military Commission. This is seen by many as an indication of President Xi Jinping's willingness to look beyond the PLA for military guidance. PLAAF is now being considered as a truly independent service capable of conducting autonomous air campaigns with other forces in support roles.

In terms of its capabilities, PLAAF has made progress in the last two decades. From a fleet of mostly obsolete 2nd and 3rd generation Soviet era aircraft, it has evolved to a modern 4th generation force featuring wide ranging capabilities. This has been possible through adopted technologies backed by a robust domestic defence manufacturing environment.

With increase in capability and lethality, the IAF also has come a long way. Having fought numerous wars, IAF is a battle-hardened and a well-trained air force. Recent conflicts have also provided IAF the institutional experience to conduct fast-paced operations in a highly networked combat environment. Building on achievements during the Kargil conflict, it has become one of the few air forces in the world that has the experience and expertise in high altitude targeting. The IAF routinely exercises with other modern air forces, has good exposure to contemporary tactics and has training standards that match the best in the world. Op advantages accrued due terrain and training will undoubtedly give the IAF an edge counterbalancing the numerical superiority of the PLAAF.

"China: The Rising Aerospace Power - Implications for India" deals with a very topical issue. I am happy that the Air Marshal has chosen to express his views on a subject that is of immediate relevance to India. He is a prolific writer who has published numerous articles on air power and strategy. The layout of chapters and detailed contents in the book are indicative of his deep research. It is apparent that the Air Marshal's wealth of experience as a Fighter Pilot and as an Experimental Test Pilot has helped him bring forth valuable insights into the technological and operational facets of air power. I strongly recommend this easy to comprehend volume to students of regional security and aerospace power.

Air Chief Marshal RKS Bhadauria,
PVSM, AVSM, VM, ADC
Chief of the Air Staff

Director's Overview

China's aerospace growth is already making major waves and the development heralds a new era in the Chinese national security approach. While Chinese aerospace power presents substantial new insight on that particular question, its focus is decidedly broader in scope. This book offers a comprehensive survey of Chinese aerospace developments, with concentration on areas previously unexplored of potential strategic significance. It also links these developments to the vast Chinese geographical theatres and battle-space of the larger Indo-Pacific region and highlights the consequent implications for the world and India in particular. A set of aerospace systems, from manned aircraft, microsatellites, unmanned aerial vehicles, ballistic and Cruise missiles are already challenging the U.S. dominance and threatening Indian security in the region. Cumulatively, progress in all major aerospace dimensions by various elements of the People's Liberation Army Air Force (PLAAF) and People's Liberation Army Navy (PLAN) signifies a new period in which Chinese forces are now decisively altering the complexion of the military balance in the region.

While many articles and books have been written on Chinese aerospace development and many more discuss future Chinese air strategy in the Indo-Pacific region, no book connects the two issues, simultaneously evaluating the Chinese aerospace challenge and its implications and options for India.

This book, "*China's Rising Aerospace Power: Implications for India*" offers both a broad strategic context for the lay reader and considerable insights for even the most well-informed specialists, with many chapters covering the growth of Chinese aerospace power and the major structural changes of the armed forces.

While technologically, Chinese aerospace power still lags considerably behind the much more advanced and powerful USA, the cash surplus with China is allowing to make significant investments in niche technologies such as stealth, artificial intelligence, hypersonic aerial systems, directed energy weapons, cyber, space and unmanned systems. In some areas, China is already about to become the world leader. The structural changes of the armed forces have made them "lean and mean" and Chinese power is already proving a deterrent while adopting a unilateral and belligerent approach in the South China Sea. Similarly, aerospace power is clearly a significant element of China's deterrence on the contested borders and maritime space and in mobilization of 'Rapid Reaction Forces or Trans-theatre Support Operation. This book covers all domains of Chinese aerospace power and gives an insight into the strengths and weaknesses while suggesting options for India.

This book relates to understanding the growing importance of aerospace power in modern warfare, including integration of air and space which has truly become the new theatre of war. The unifying space dimension will remain the single most important source for information and communication which can be used in multiple forms. It attempts to focus on China's evolving military strategy with emphasis on the aerospace dimension and its implications in the region, specifically in the Indian context. It also attempts to analyse the strategic importance of the rising economic, political, and military stature of China to understand its regional and global implications in the military domain. China, in recent times, has repeatedly demonstrated its resolve to defend its security interests at all costs. For over three decades, it has been modernising its military capabilities besides developing new and complex military platforms that would be of great value to prosecution of joint operations warfare. These are critical times for all the People' Liberation Army air arms which attempt to integrate many new and complex platforms, and adopt modern operational concepts, including network-centric warfare. Besides undertaking a comprehensive modernisation of its armed forces, China is developing a series of offensive space capabilities while advocating the peaceful use of outer space. This book highlights the transformation of China's aerospace power. The PLAAF appears to be well on its way to becoming a modern, highly capable air force for the 21st century. China's latest combat aircraft, the J-20, highlights China's ambition to produce a fighter aircraft that incorporates stealth attributes, advanced avionics, and super-cruise capability. China's aviation industry is in parallel developing all genres of platforms. The PLAAF exercises have demonstrated that its role has changed from support to ground forces to being able to conduct operations independently. It is actively trying to imbibe better training programmes and has increased joint training with other arms and other air forces in recent years.

China is likely to depend heavily on surface-to-surface missiles, as they are more difficult to defend against. Though their territorial claims are more aggressive in India's Arunachal Pradesh, they have been no less active in the Ladakh region; national highway 219 passes through Aksai Chin that is illegally occupied by China. China is also likely to use force multipliers such as AWACS and Flight Refuelling Aircraft for its air-campaign against India.

As part of the options for India, this book recommends full national effort for indigenisation of production of aircraft to rapidly achieve the desired force levels. It also book emphasizes the need for India to invest in key aerospace technologies. It recommends diplomatic initiatives and building alliances in the region.

With a large quantum of current factual data, this book will be of immense value for readers of the Asian region as well as the strategic community across the globe. I wish to place on record my gratitude to the author for undertaking this project of vital national security importance for the USI of India. I also compliment the author for producing a very well-researched book.

Maj Gen B K Sharma AVSM, SM**
Director, United Service Institution of India

Preface

"China is a sleeping giant. Let her sleep, for when she wakes she will move the world."

—Napoleon Bonaparte

Rapid economic growth in the last two decades has allowed China to spend large sums on military technologies and modernisation. Its military is thus well on its way to bridging the aviation technology gap with the USA. China took the shortcut to achieve this unsurpassed achievement by reverse engineering Soviet and later Russian aircraft and engine designs. In more recent years, it allowed high technology U.S. companies to set up shop in China and in turn acquired some high-end technologies. China is also often accused of using the Chinese Diaspora in the USA and cyber-attacks to steal sensitive technologies, thus circumventing the long tedious route of new platform and technology evolution. With the breakdown of the Intermediate-Range Nuclear Forces (INF) Treaty, there is once again a race for long-range weapons. China is today developing world-beating advanced weapon systems, and concentrating on developing long-range ballistic and Cruise missiles to cease advantage in the long-distance, well Beyond Visual Range (BVR) battle. China has traditionally relied on its large mobile, surface-based missile strike capability vis-a-vis American concentration on a potent stealth bomber force. China accordingly created an independent People's Liberation Army Rocket Force (PLARF), and has been displaying its hypersonic missiles and the 'carrier killer', 1,500 kilometre range DF-21D. The Chinese already claim to be ahead of the USA on this count.

Chinese armed forces have been going through three types of equipment transformation; modernising and replacing the legacy equipment across all services and branches. The 2000 Defence White Paper stated that the PLA should move to informatised systems, roughly analogous to the concept of net-centric capability using advanced IT and communications systems to gain operational advantage. Space, cyber, and electromagnetic warfare capacities have the potential to paralyse even a high-tech system of the enemy. Also, information-enabled capabilities and boost command, control and communications. The White Paper talks of 'intelligent warfare is on the horizon'. It involves military systems that will be enhanced through integration of advanced automation, big data, and artificial intelligence (AI). Intelligent processing of data collected from reconnaissance, surveillance, or intelligence systems. They have been talking of human–machine intelligence, data-centric analysis, independent decision-making and autonomous-attack capabilities. The Chinese government has aggressively launched 'Made in China 2025' project to acquire and master all high technologies and manufacturing capabilities. They plan to select core state-owned industries, centrally inject funds, and emphasise on intellectual property acquisition.

China is already called the 'Walmart' of small unmanned aerial vehicles (UAV) and the largest producer of hand-held systems. Fifty per cent of the world's UAVs for hobbyists are made in China. It is rapidly advancing in AI based autonomous unmanned systems, and intelligent robots. AI is also being used for hybrid intelligence, and automated decision making. China is also concentrating on finding means to exploit vulnerabilities to degrade or jam an adversary's communication systems to prevent their ability to launch offensives in a network centric operational environment. They have invested heavily in battlefield disruptive technologies, including artificial intelligence, hypersonic systems, cyber-warfare, long-range weapons and advanced space based systems. China is also investing heavily in directed energy weapons research and has the second-largest economy in the world, which is predicted to overtake the USA's GDP by 2030. It is targeting to have all top-end technologies by 2025, and surpass the USA as a military power by 2040.

The People's Liberation Army Navy (PLAN) has a significant air arm with nearly 500 aircraft. It plans to build a significant force of up to six aircraft carriers by 2035. Currently, it has only two, including the first indigenous Type 001A carrier. The PLAN is, however, rapidly producing smaller warships 'corvettes' and 'frigates', designed with an eye to "littoral" combat or fighting close to China's shores. The Chinese aircraft carrier program, including the development, production and operation is primarily driven by PLAN. Since 1985, China will at any time have one Carrier Group in the SCS and one in the Indian Ocean. The PLA operates close to 600 helicopters including heavy lift, medium utility cargo, and attack helicopters. The helicopter types are Z-9, Z-19, Mi-17, Sikorsky S-70 and Eurocopter EC 120. The PLA also has a few small transport aircraft for communication duties.

Taking a leaf from the American approach and thinking, China realised very early that the one who controls the aerospace domain, controls the planet. It has thus invested heavily in air and space systems. These include space based platforms, stealth fighters, heavy transport aircraft, helicopters and UAVs. China is investing in many aircraft carriers with a desire to dominate at least the eastern Pacific Ocean and parts of the Indian Ocean. Both these are critical to the Chinese sphere of influence.

China's authoritarian central control helps quicker decision making. China's military muscle flexing has been seen in the manner it tries to bulldoze while trying to resolve border disputes with its neighbours. It annexed Tibet when it was not even a significant military power. Chinese aggressive foray and postures in the South China Sea found the world helplessly watching as it usurped millions of kilometres of resource rich and strategically important sea. Unlike the USA that was initially spending huge amounts during the Cold War in defence of Europe, and later bogged down in the oil rich Middle East, China has been relatively less involved and has been quietly busy building its economy and military capabilities. Meanwhile, cash rich China also realised that a better way of increasing influence was to financially help under-developed, relatively poor, nations by offering debt financing of important infrastructure projects. This not only gave a foothold into these countries but also created markets for its excess manufacturing capacity and surplus produce after shrinking European markets, and the American trade embargoes. The Belt and Road Initiative (BRI) has now extended to most of the Eurasian land mass. While there are debt-trap related issues, Chinese influence has actually spread to the detriment of both the USA and India. China's heavy investments in the China-Pakistan Economic Corridor (CPEC), some of which are strategic in nature, are of direct concern to India. Similarly, its investments in other Indian neighbours such as Myanmar, Bangladesh and Sri Lanka, among others, could result in strategic encirclement.

In the last two decades, China has made great strides in developing airpower capability, albeit, it remains work-still-in-progress. The introduction of fifth-generation fighters, bombers, unmanned aircraft and advanced Cruise missiles has transformed it from a defensive force to one that can project Chinese power throughout Asia and its eastern Pacific sea board. China is pulling far ahead in military capability vis-a-vis India and the gap is continuously increasing. China is also playing a dominating role in most world bodies like the United Nations (UN) and Shanghai Cooperation Organisation (SCO).

China's space program was partly led by the initial thrust on ballistic missiles. Currently China has around 260 ICBMs. They put a human in space in 2003, and sent an un-crewed moon explorer in 2007. China also conducted a successful Anti-Satellite (ASAT) test in 2007. All elements of the 100-tonne Chinese permanent space station are in place and it should become operational by 2021. The 33 satellite BeiDou Navigation Satellite System is providing global services since December 27, 2018. Two more satellite have been launched to complete the third generation global system. It is claimed that BeiDou-3 will reach millimetre-level accuracy, with post-processing. A human expedition to Moon is on the cards around 2025. Missions to Mars and beyond are also planned. China wants to create energy from space through an orbital solar power plant. China is a signatory of all space treaties for peaceful use. It is also conceptualizing a crewed lunar base and studying a possible space based ASAT capability.

The defence of the mainland remains the primary focus of Chinese military doctrine and training. China's immediate strategic target is to reclaim the Republic of China (ROC) Taiwan. They are also very suspicious of the militarily re-emerging Japan with whom they have historic scores to settle. Yet, India is their real long term challenger in the Asian region. India has to study Chinese strategy and designs closely. In a conflict with India they plan to push the forward edge of battle into Indian territory. They place higher reliance on the use of surface-to-surface missiles (SSM), which they feel are difficult to defend against. In the absence of effective radars along the mountainous border, they will make extensive use of Airborne Warning and Control Systems (AWACS). The high altitude airfields in Tibet would place a penalty on weapon load, and would therefore necessitate aerial refuelling. China's own secure satellite navigation system, BeiDou, would be very handy for navigation and accurate targeting. Having annexed significant parts of India's Ladakh in the 1962 Sino-Indian war, their focus is likely to be in Arunachal Pradesh. India is still in the process of strengthening air infrastructure in the east. Meanwhile, China is conscious of lack of low combat experience of its crew and has started more air exercises with other air forces. Also, it has had a total relook at its aviation training.

Of immediate concern to India are the Chinese border incursions. As China still builds military strength, it does not plan a war but is resorting to offensive posturing. It is using its growing military might to send deterrent signals. However, in the last two decades, China has built a significant road and rail network in the Tibetan region. It has laid oil pipelines with the dual aim of regional economic development and military logistics capability. China already has eight fully developed dual-use airfields and three more are under construction. Most of these airfields are at altitudes around 3,500 to 4,500 metres above mean sea level (AMSL). Bangda airport in southwest Tibet Autonomous Region has an elevation of 4,334 metres AMSL and Daocheng Yading airport under construction is 4,411 metres high. The PLAAF is conscious of this and has built longer runways and factored in the need for Flight Refuelling Aircraft (FRA). On this count, India does have an advantage with much larger numbers of airfields and at much lower altitudes. The PLAAF can also fly from some lower altitude airfields east of Myanmar such as Kunming, but that could mean overflying Myanmar, with or without its permission. China is likely to make extensive use of SSMs. India needs to closely study, identify and exploit weaknesses in China's strategy.

With Pakistan and China as two-nuclear neighbours, both of which are strategically allied, have some commonality in equipment, exercise together, and have boundary disputes with India, there is the need for India to factor in a possible two-front war. Various scenarios need to be evaluated for such collusive action. While India is gradually strengthening military and civil infrastructure in its north east, a lot more is still work in progress. For the air war, the IAF needs to first build back to the authorised 42 combat squadrons from the current all time low of 30. For a two-front war against the second and seventh largest air forces of the world, some analysts feel that around 50 combat squadrons would be desirable. Certainly, India needs to invest in more SSMs, Cruise missiles and Surface-to-Air Missiles (SAM). The Indian Army will require heli-lift support for inter valley movement. India would need many more AWACS and high altitude UAVs. India needs to train much more in the mountains. For India to become a significant world power, and taking a leaf from China, India needs to invest much more human and financial defence capital in Research and Development (R&D), and critical technologies. India must work towards weakening China's ability to drive its narrative.

Over the years, many books have been written on Chinese air power and its implications for India. Both India and China are rising economies. Both countries are growing militarily. Also, the military technologies are evolving rapidly and the way wars are fought is gradually changing. Air and space as mediums are tending to merge with more and more platforms transiting through them for operational reasons. Because of increasing importance, aerospace is getting the highest capital budget funding in most countries. Space has not only become a great force multiplier, it could even become a major war zone. The future is in AI-based unmanned systems and robotics. There are many other disruptive technologies unfolding that will change the nature of warfare. Both China and India are not only inducting new state-of-the-art systems but also relooking at the command and control structures to make the military lean-and-mean.

This book is aimed at increasing understanding of the growing Chinese air and space power and how the aerospace has become the new theatre of war. It looks at how the PLAAF is fast emerging from the shadows of the Chinese Army, and trying to imbibe modern technologies and strategy to use modern platforms. It also tries to suggest how the IAF must prepare to face such a challenge in case of an air war across the Himalayas.

Air Marshal Anil Chopra, (Retd),
PVSM, AVSM, VM, VSM

ACRONYMS

A2/AD	Anti Access/Area Denial
AAM	Air-to-Air Missiles
ABM	Anti-Ballistic Missile
ACMI	Air Combat Manoeuvring Instrumentation
AD	Air Defence
ADIZ	Air Defence Identification Zone
AESA	Active Electronically Scanned radar Array
AEW	Airborne Early Warning
AEW&C	Airborne Early Warning and Control
AFNET	Air Force Network
AGM	Air-to-Ground Missile
AI	Artificial Intelligence
AIFTA	ASEAN–India Free Trade Area
AIIB	Asia Infrastructure Investment Bank
ALCM	Air Launched Cruise Missiles
ALG	Advance Landing Ground
AMCA	Advanced Medium Combat Aircraft
AMSL	Above Mean Seal Level (AMSL)
AON	Acceptance of Necessity
ARM	Anti Radiation Missiles
ASAT	Anti-Satellite
ASW	Anti Submarine Warfare
ATGM	Anti-Tank Guided Missiles (ATGM)
AVIC	Aviation Industry Corporation of China

AWACS	Airborne Warning And Control Systems
AWS	Air Weapons School
BAI	Battlefield Air interdiction
Bde	Brigade
BDS	BeiDou Navigation Satellite System
BECA	Basic Exchange and Cooperation Agreement
BIMSTEC	Bay of Bengal Initiative for Multi-Sectoral Technical and Economic Cooperation
BMD	Ballistic Missile Defence
BOP	Border Out Post
BRI	Belt and Road Initiative
BSM	Battlefield Support Missile
BVR	Beyond Visual Range
C2	Command & Control
CAATSA	Countering America's Adversaries Through Sanctions Act
CAC	Chengdu Aerospace Corporation
CAIC	Changhe Aircraft Industries Corporation
CAO	Counter Air Operations
CAR	Central Acquisition Radar
CASC	Chinese Aerospace Science and Technology Corporation
CASIC	Chinese Aerospace Science and Industry Corporation
CASTIC	China Aerospace Science and Technology Corporation
CATOBAR	Catapult Assisted Take-Off But Arrested Recovery
CENTO	Central Treaty Organization
CEO	Chief Executive Officer
CISMOA	Communications and Information Security Memorandum of Agreement
CMC	Central Military Commission
CMI	Civil Military Integration
COMAC	Commercial Aircraft Corporation of China
COMCASA	Communications Compatibility and security Agreement
COTS	Commercial-off-the-shelf

CPEC	China-Pakistan Economic Corridor
CPTPP	Comprehensive and Progressive Agreement for Trans-Pacific Partnership
CSFO	Counter Surface Force Operations
DAMPE	Dark Matter Particle Explorer
DBO	Daulat Beg Oldie
DCA	Defence Cyber Agency
DEW	Directed Energy Weapons
DRDO	Defence Research and Development Organisation
DRFM	Digital radiofrequency memory
DSA	Defence Space Agency
EBO	Effect Based Operations
EEZ	Exclusive Economic Zone
ELINT	Electronic intelligence
EM	Electro Magnetic
EMALS	Electro Magnetic Aircraft Launch System
EO	Electro-Optical
EW	Electronic Warfare
FAWC	Foreign Affairs Work Conference
FEBA	Forward Edge of Battle Area
FET	Field Evaluation Trial
FGA	Fighter/Ground Attack
FGFA	Fifth Generation Fighter Aircraft
FOC	Final Operational Clearance
FRA	Flight Refuelling Aircraft
GaN	Gallium Nitride
GOE	Government Owned Enterprises
GSOMIA	General Security Of Military Information Agreement
HADR	Humanitarian Assistance and Disaster Relief
HMS	Helmet-Mounted Sight
HOTAS	Hands-On-Throttle-And-Stick
HUD	Heads-Up Display
IA	Indian Army
IAF	Indian Air Force
ICBM	Inter Continental Ballistic Missiles

ICO	Information-Centric Operations
IFF	Identification, Friend or Foe
INDOPACOM	Indo-Pacific Command
INF	Intermediate-Range Nuclear Forces
IOC	Initial Operational Clearance
IPCOM	US Indo-Pacific Command
IS	Islamic State
ISR	Intelligence Surveillance and Reconnaissance
ISRO	Indian Space Research Organisation
ITBP	Indo-Tibetan Border Police Force
LAC	Line of Actual Control
LCA	Light Combat Aircraft
LDP	Laser designator pods
LEMOA	Logistics Exchange Memorandum of Agreement
LIFT	Lead-in fighter trainer
LRDE	Electronic & Radar Development Establishment
LRTR	Long Range Tracking Radar
MAFI	Modernization of Airfield Infrastructure (MAFI)
MALE	Medium Altitude Long Endurance
MFD	Multi Function Displays
MIRV	Multiple Independent Re-entry Vehicle
MPR	Medium Powered Radar
MRAF	Military Region Air Force
MWF	Medium Weight Fighter
NASAMS	National Advanced Surface-to-Air Missile System
NATO	North Atlantic Treaty Organization
NCR	National Capital Region
NSG	Nuclear Suppliers Group
OBOR	One Belt, One Road
OODA	Observe, Orient, Decide, Act
ORP	Operational Readiness Platform
PAC	Pakistan Aeronautical Complex
PACAF	Pacific Air Forces
PACOM	US Pacific Command
PAF	Pakistan Air Force

PAT	Passive Aero-elastic Tailored
PDI	Pacific Deterrence Initiative
PGM	Precision Guided Munitions
PLA	People's Liberation Army
PLAAF	People's Liberation Army Air Force
PLAN	People's Liberation Army Navy
PLANAF	People's Liberation Army Naval Air Force
PLARF	People's Liberation Army Rocket Force
PLASOF	PLA Special Operations Forces
PLASSF	People's Liberation Army Strategic Support Force
PNT	Positioning, Navigation, and Timing
PoK	Pakistan occupied Kashmir
PPP	Pakistan People's Party
PRC	People's Republic of China
Quad	Quadrilateral Security Dialogue (QSD)
QUESS	Quantum Experiments at Space Scale
R&D	Research and Development
RAF	Royal Air Force
RCEP	Regional Comprehensive Economic Partnership
RCS	Radar Cross Section
Regt	Regiment
RFI	Request for Information
RFP	Request for Proposal
RIMPAC	Rim of the Pacific
ROC	Republic of China
ROK	Republic of Korea
SAARC	South Asian Association for Regional Cooperation
SAM	Surface-to-Air Missile
SAR	Search and Rescue
SASTIND	State Council's State Administration for Science, Technology, and Industry for National Defence
SCO	Shanghai Cooperation Organisation
SCS	South China Sea
SEAD	Suppression of Enemy AD
SEATO	Southeast Asia Treaty Organization

SIGINT	Signal Intelligence
SOF	Special Operations Forces
SRBM	Short Range Ballistic Missile
SSM	Surface-to-Surface Missiles (SSM)
STOBAR	Short Take-Off But Arrested Recovery (STOBAR)
TAR	Tibet Autonomous Region
TCAF	Theatre Command Air Force
THAAD	Terminal High Altitude Area Defence
TOW	Tube-launched, Optically tracked, Wire-guided
TPP	Trans-Pacific Partnership
TRCC	Turbine Rocket Combined Cycle
TRRE	Turbo-aided rocket-augmented ram/scramjet engine
UAS	Uninhabited Aerial Systems
UAV	Unmanned Aerial Vehicles
UN	United Nations
UNGA	UN General assembly
UNSC	United Nations Security Council
USAF	U.S. Air Force
USN	U.S. Navy
VASIMR	Variable Specific Impulse Magneto-plasma Rocket
WSO	Weapon Systems Operator
WTC	Western Theatre Command
XR-SAM	eXtra Long Range Surface to Air Missile

1

Introduction
China: The Rising Aerospace Power

"To secure peace is to prepare for war".
—**Carl von Clausewitz**

Napoleon once said, "China is a sleeping giant, let her lie and sleep, for when she awakens, she will astonish the world."[1] By the turn of the 21st century the sleeping giant was not only awake but heavily influencing and, in many ways dictating world affairs. The People's Republic of China (PRC) has had an unprecedented continuous double-digit growth of its economy for nearly two decades, and is still maintaining close to 7 per cent growth. In the same period, Europe and the USA who have been preoccupied with the 'war on terror', have had declining or slow rates of growth. While China was raising the standard of living of its people, it has simultaneously been modernizing its armed forces. The balance of world power has already shifted from Europe and the Atlantic to the Indo-Pacific region. Many in the world and, in particular the West, have been seeing a rising China as a disruption of the status quo of power balance that had evolved after World War II. But China, while strengthening its political, economic and military strength, has been referring to it as "China's peaceful rise". During the leadership of Hu Jintao (2002-12), General Secretary of the Communist Party of China, the term "China's peaceful development" was coined to rebut those who were scaring the world with the "China threat theory".[2] China also claimed that historically it was not an aggressive empire. It wanted to be seen as a responsible world leader who did not interfere in the affairs of others and wanted to avoid a confrontation. The former vice-principal of the Central Party School, Zheng Bijian, in his speech in late 2003 during the Boao Forum for Asia, had pointed out that in the past, a rise of a new power often resulted in drastic changes to global political structures, and even war.[3] He believed that this was because these powers chose the road of aggression and expansion, which will ultimately fail. Zheng stated that the PRC should instead develop peaceably, and in turn, help to maintain a peaceful international environment. China's rapid development was viewed by them as a multi-polarizing change that challenged the uni-polarity of the world structure under the USA.

Consolidated Power and Push

In his speech on November 29, 2014 to the Foreign Affairs Work Conference (FAWC)—the first to be held since 2006—Chinese President Xi Jinping exhorted the assembled Chinese officials to develop for China a 'distinctive diplomatic approach befitting its role as a major country', stressing that China must 'conduct diplomacy with a salient Chinese feature and a Chinese vision'.[4] China sought a new international political and economic order, but it was one that can be achieved through incremental reforms and the democratization of international relations. China would "not follow the path of Germany leading up to World War I or those of Germany and Japan leading up to World War II, when these countries violently plundered resources and pursued hegemony. Neither will China follow the path of the great powers vying for global domination during the Cold War."[5] The U.S.A. in fact helped China to become a privileged member and shaper of the international system. As China grows more powerful, it is displacing decades-old American pre-eminence in parts of Asia. The outlines of the rivalry are defining the future of the continent.[6] On Taiwan, the PRC policy since 2005 has been determined by the Anti-Secession Law. The essence of Article 8 of the law is that Taiwan should be united with mainland China, preferably by peaceful means; but if Taiwan formally declares independence or refuses to submit, the PRC will use force.[7]

Peaceful Development Strategy

In 2005, the State Council of the PRC issued a White Paper defining China's peaceful development strategy.[8] Its five chapters essentially covered economic development according to global norms; desire for a multi-polar world rather than hegemony; poverty reduction, and reducing energy consumption; develop science for domestic market growth and cleaner industrialization: more use of information technology; innovative exploitation of human capital through education; support regional integration; invest abroad and maintain large labour force and exports for use abroad; promote democracy in international relations; pursue arms control and nuclear disarmament and resolve remaining border disputes peacefully. The ground reality has been somewhat different till now. China remains assertive in most of its many border disputes.

Ana Nicolaci da Costa, BBC business reporter, wrote on how the world is grappling with China's rising power.[9] "China's sheer size and population make it a heavyweight, and a clear strategic rival to the United States. Its influence has boomed—along with its economy—in recent years, as the US and Europe nursed the wounds from devastating financial crises," she wrote. The USA was keen to retain its dominant position. US President Donald Trump's Administration launched a trade war with China in 2018, hitting about half of Chinese imports into the USA through tariffs. The tariffs were a response to China's "unfair" trade practices and alleged intellectual property theft. Many in Beijing suspected that the USA wanted to block China's rise—which is seen as a challenge to the USA's own established hegemony. Speaking at the World Economic Forum in 2017, Chinese President Xi Jinping said: "Countries have the right to development, but they should view their own interests in the broader context. And refrain from pursuing their own interests at the expense of others."[10] In a recent speech, US Vice-President Mike Pence said China had chosen "economic aggression" when engaging with the world and "debt diplomacy" to spread its influence, he added. "The Chinese think the USA wants to contain them and certainly a lot of people here do. A lot of people in the USA think the Chinese want to take over the world," said C. Fred Bergsten, founding director of the Peterson Institute for International Economics in Washington.[11] Meanwhile, Xi Jinping's greater claims to power are making many nervous. "The big egos and strong stances of the two leaders are exacerbating it, so at the moment it's really a collision course towards a cold war," said Mr Bergsten, who previously also worked as former US Secretary of State Henry Kissinger's deputy for economic policy.[12]

Chinese BRI and Telecom Concerns

National security worries have also led to curbs on Chinese companies, such as telecom giants Huawei and ZTE, and on Chinese investments abroad. Countries that are supposed to be benefiting from China's increased wealth also seem to be growing more cautious. The BRI, unveiling since 2013, aims to expand trade links between Asia, Africa, Europe and beyond. But the multi-billion dollar project, which is causing debt concerns in certain countries like Sri Lanka, Malaysia and Pakistan are facing growing resistance as recipients fear debt accumulation and increased Chinese interference in domestic politics and influence. "I think first and foremost this is a tool for China to expand, to strengthen its soft power influence through economic diplomacy," said Michael Hirson, Asia director at Eurasia Group.[13] "There is also a strong strategic dimension which comes into play in the projects that focus on the energy sector and on port deals which serve China's interest in securing strategic assets overseas," said US Vice-President Mike Pence flagging the experience of Sri Lanka, which had to hand over control of a port to China to help repay foreign loans. "China is offering hundreds of billions of dollars in infrastructure loans to governments from Asia to Africa to Europe and even Latin America," Mr. Pence said.[14] The terms of those loans are opaque at best, and the benefits invariably flow overwhelmingly to Beijing.

The Unpredictable Rise and Concerns

In his paper, 'The Rise of China and the Future of the West. Can the Liberal System Survive?' G. John Ikenberry[15] argues that "the rise of China will undoubtedly be one of the great dramas of the twenty-first century. China's extraordinary economic growth and active diplomacy are already transforming East Asia, and future decades will see even greater increases in Chinese power and influence. But exactly how this drama will play out is an open question. Will China overthrow the existing order or become a part of it? And what, if anything, can the United States do to maintain its position as China rises?" Some observers believe that the American era is coming to an end, as the Western-oriented world order is replaced by one increasingly dominated by the East. Realists feel that as China gets more powerful and the position of the USA erodes, two things are likely to happen: China will try to use its growing influence to reshape the rules and institutions of the international system to better serve its interests, and other states will start to see China as a growing security threat. The result of these developments, they predict, will be tension, distrust, and conflict, the typical features of a power transition. While America seeks a uni-polar world with bi-polar Asia, China clearly wants a bi-polar world and uni-polar Asia.

Daniel Blumenthal, director of Asian Studies at the American Enterprise Institute, writes in his article, 'The Unpredictable Rise of China' that Xi Jinping seeks national rejuvenation, but his nation's mounting power masks increased instability.[16] Since the end of the Cold War, Beijing has viewed Washington as its chief geopolitical rival, yet official Washington has only recently awakened to this strategic competition. But as American observers start to see China's ambitions more clearly, they have also begun to misdiagnose the challenges they pose. "Political scientists are discussing the 'power-transition theory' and the 'Thucydides trap' (when one great power threatens to displace another, war is almost always the result—but it doesn't have to be), as if China were on the verge of eclipsing the USA in wealth and power, displacing it on the world stage," he said. According to him, two things emerge. The first is that this is not how the Chinese themselves understand their rise. When Chinese President Xi Jinping calls for the Chinese to realize the "China dream of national rejuvenation," he is articulating the belief that China is simply reclaiming its natural political and cultural importance. China is not, as was once said of imperial Germany after its unification, "seeking its place in the sun." Rather, it is retaking its rightful place as the sun. The second is that it's an open question whether China will achieve rejuvenation in the face of both a seemingly stagnating economy and party factionalism. Xi is more powerful than his predecessors, but his rule is also more fragile. The Chinese Communist Party (CCP) has long faced a crisis of legitimacy, but Xi's transformation of China into a high-tech police state might hasten this crisis. These factors combine to make China more dangerous in the short term but also less competitive in the longer term. This means that the People's Republic of China perceives an opportunity for "great renewal" even as it will be less powerful than was expected.

Xi Jinping's Personal Power Play

Jonathan Tepperman, editor-in-chief of *Foreign Policy*, in his article 'China's Great Leap Backwards' writes, "For decades, the country managed to avoid most problems suffered by dictatorships.[17] Now Xi Jinping's personal power play risks undermining everything that made China exceptional. In the last 40 years, China has seen unprecedented economic growth, lifting 800 million people out of poverty. China scholar Orville Schell describes this record as "one of the most startling miracles of economic development in world history." "Under the guise of fighting corruption, President Xi Jinping is methodically dismantling and is erecting a colossal cult of personality focused on him alone, concentrating more power in his hands than has any Chinese leader since Mao Zedong," Jonathan says. To maintain power, some individuals repress dissent and rule by intimidation. Because bureaucrats and citizens live in fear, they compete to flatter their bosses. Nobody tells the truth, especially when it could make them or their leaders look bad. As a result, cloistered tyrants—their egos bloated by constant, obsequious praise—find themselves increasingly cut off from reality and the rest of the world. Until Xi assumed power in 2012, China was building what scholars have called an 'adaptive authoritarian' regime. While remaining nominally communist, the country embraced many forms of market capitalism and a number of other liberalizing reforms. Censorship never disappeared, but party members could disagree and debate ideas, and internal reports could be surprisingly blunt. No longer. Today, Xi is systematically undermining virtually every feature that made China so distinct and helped it work so well in the past. He is undoing the most unusual feature of the system Deng created which allowed distributed power among various leaders. Since assuming power in 2012, Xi has worked to dismantle China's collective leadership system in several ways. First, in the name of fighting corruption—an important goal and one that China badly needs—he has purged a vast number of officials whose real crime, in Xi's view, was failing to show sufficient loyalty to the paramount leader. Xi has also consolidated his power by abandoning the term limits on his job and by refusing to name a successor, as his predecessors did halfway through their tenures. He has also assumed direct control of the armed forces; and made himself 'Chairman of everything' and a large number of working groups on policies ranging from finance to Taiwan to cyber-security, report directly to him. The odds that China will experience a seriously destabilizing economic crisis is what Ruchir Sharma, the head of emerging markets at Morgan Stanley, has been predicting for years, are rising. "The big question is whether one of the ticking time bombs—bad debt, overheated property markets, oversized state owned enterprises—will explode," says Alexander Gabuev, a China specialist at the Carnegie Moscow Centre. "Because of Xi's concentration of power, no one will give him advance warning if one of these bombs is about to go off." Since taking power, Xi has charted a far more aggressive foreign policy than his predecessors, alienating virtually every neighbour and the USA by pushing China's claims in the South China Sea, threatening Taiwan, and using the military to assert Beijing's claims to disputed islands.

Military Modernisation and Reorganisation

On the other hand Xi reminds that the Chinese Communist Party (CCP) had long struggled to restore China to its historic centrality in international affairs. "Ours is a great nation," he said, "that has endured untold hardships and sufferings."[18] But the Communist Party, he said, had forged ahead, "thus opening a completely new horizon for the great renewal of the Chinese nation." The People's Liberation Army (PLA) is developing its capabilities at a rapid speed, changing the balance of power in Asia to its advantage. The International Institute for Strategic Studies estimates that since 2014 the People's Liberation Army Navy has "launched more submarines, warships, principal amphibious vessels and auxiliaries than the total number of ships currently serving in the navies of Germany, India, Spain, Taiwan and the United Kingdom."[19] Its shipbuilding program is outpacing that of the USA. Between 2016 and 2017, 32 new ships were commissioned by the PLA, in comparison to 13 by the USA. China is also spending vast sums on breakthrough technologies like artificial intelligence, hypersonic systems and robotics, which could tilt the nature of warfare to its advantage. Ben Westcott of CNN in his analysis, states that "China's military is going from strength to strength under Xi with a focus on fighting and winning future wars."[20] Xi has also embarked on a massive internal reorganization of the PLA, streamlining the organization and bringing it firmly under his control. China's 2019 defence budget of US$ 177.54 billion grew 7.5 per cent over previous year, albeit as it is still very small vis-a-vis the US defence budget of US$ 750 billion.[21] China's military modernisation objectives remain as ambitious as ever. It continues to plan for multiple contingencies. While the US Navy still dominates the world's oceans, People's Liberation Army Navy (PLAN) is rapidly gaining ground by churning out naval vessels and making technological advances, according to Andrew Erickson, professor of strategy at the US Naval War College's China Maritime Studies Institute. "No one has presided over this level of Chinese military development in Chinese history before Xi Jinping," he said.[22]

Xi has also begun a reorganization of military personnel, stressing quality and effectiveness over quantity of staff. Many former generals have been disciplined or imprisoned for corruption in an attempt to professionalize the armed services. Despite Western concerns, Zhang Yesui, the National People's Congress spokesman, recently told reporters: "China's limited defence spending, which is for safeguarding its national sovereignty, security and territorial integrity, poses no threat to any other country."[23] The Chinese government has built a navy and armed forces designed to protect the country and exert its influence in the surrounding region, especially the East and South China seas. The Chinese government's focus is on controlling the South China Sea (SCS) where it has been building militarized artificial islands and holding regular drills. Pakistan owes its "all weather friend" China at least $10 billion debt for the construction of the Gwadar port and other projects, a top US general has said, as he underlined Beijing's "predatory economics" to expand its global influence, reported Lalit K. Jha in the *Live Mint*.[24] Gwadar is de facto a Chinese port with full access to PLAN. PLAN opened its first international base in Djibouti in July 2017.

Aerospace: The Doman of Focus

Aerospace has been the key area of Chinese military investments. China is conscious about the primacy of aerospace power as the major instrument of waging modern wars. China wants to harness the speed, range, lethality, accuracy and flexibility that aerospace offers. Space stations and satellites not only act as vantage points for reconnaissance and communications, but space is also used for weapon transition and future manned flights. Armies and navies around the world are investing more and more in element air. Aerospace is technology and cost driven, and requires huge capital acquisition budgets. China thus has given itself targets. 'Made in China 2025' is the target for all aerospace technologies.[25]

China has consolidated its aviation industry under the state—owned aerospace and defence umbrella organization called the Aviation Industry Corporation of China (AVIC). It was last ranked 143th in the *Fortune* Global 500 list[26] AVIC has over 100 subsidiaries, 27 listed companies and 542,236 employees across the globe. The major plants are at Chengdu, Shenyang, Xian, and Changhe. It covers manufacturing of all military, air transport, and general aviation aircraft. AVIC also manufactures helicopters, UAVs, aircraft systems and automobiles. Initial aircraft building plants in China had come up with Soviet/Russian support. As the Chinese industry began to grow, China wanted to free itself of Russian control and therefore began stripping Russian aircraft and systems and weapons to understand technology and manufacturing details and started reverse engineering systems much to the anger and helplessness of the Russians. They have often been accused of infringing Intellectual Property Rights (IPR), both by Russia and the USA. China however continues to lag in critical aviation technologies including, aircraft engines, Air Interception (AI) radars, Electronic Warfare (EW), stealth technologies, ejection seats, and also Precision Guided Munitions (PGM) technologies. But all this is work-in-progress. China is also working on Lasers and Directed Energy Weapons (DEW), Artificial Intelligence (AI) applications for defence aviation, and hypersonic systems technologies. Between 1997 and 2017, China's share of the global research and engineering budget grew 900 per cent, from 3 per cent to 27 per cent.[27] At US$ 177.6 bn, China's 2019 defence budget is over three times that of India. China was spending close to 20 per cent of its defence outlay on R&D as compared to a mere 5-6 per cent in India.[28] Amounting to nearly US$ 35 billion budgeted for defence Research and Development (R&D) every year, the Chinese are fast catching up with the USA.

Space: Frontier of the Future

China already has over 250 satellites and in 2018 it had 38 space-centric rocket launches[29] surpassing the USA which had 34. Its 'Long March' rockets are among the most powerful in the world that can take a 5,500 kg satellite into geostationary orbit and a 12,000 kg satellite into space for a Leo orbit. The newer variant Long March 5B, which is 53.7 metres long and 849 metric tons in weight, was launched in May 2020.[30] Space is also corporate China's newest frontier with 80 space start-ups. The BeiDou satellite navigation system consists of 35 satellites and has begun providing global services.[31] BeiDou is providing an alternative global navigation satellite system to the United States-owned Global Positioning System (GPS) and is expected to be more accurate. China has a clear time table for its Moon and Mars Lander. China has around 260 Intercontinental Ballistic Missiles (ICBM). They put humans in space in 2003 and an un-crewed moon explorer in 2007. Their target is to have a 60-tonne permanent space station by end 2020, many elements of which are already in place. A human expedition to the Moon is targeted for 2025. Missions to Mars and beyond are planned. China also plans an Orbital Solar Power Plant to tap energy from space. China also has a plan for a crewed lunar base. China is a signatory of all treaties for peaceful use of space. They had carried out a successful anti-satellite weapon (ASAT) test in 2007 when they blew up an old defunct Chinese weather satellite. More tests have since been carried out to fine tune what the Chinese call 'ground-based midcourse missile interception technology'. Some analysts believe that the Chinese are simultaneously trying to perfect technology against ballistic targets. China is also studying possible space based ASAT systems.

PLAAF

The Chinese air force, the People's Liberation Army Air Force (PLAAF), has also been regularly inducting new and improved planes and weapons, including the twin-engine J-20 stealth fighter. PLAAF is now the third largest in the world, and is closing the gap with the USA across a spectrum of capabilities. China's combat aircraft building program includes two stealth fifth-generation fighters (J-20 and J-31), a stealth bomber (H-20), a large transport aircraft (Y-20), an AWACS, variety of utility and attack helicopters and Unmanned Aerial Vehicles (UAV). China has a 'Wing Long' Medium Altitude Long Endurance (MALE) UAV program and their own cruise missiles up to 1500 kilometre range. China has even shown an interest to revive the world's largest cargo aircraft, the Ukraine's 640-tonne Antonov An-125 cargo aircraft which was halted in 1994 for lack of funding.[32] China is also designing its own 'Mother of All Bombs' to be dropped one day from the H-6K bomber.[33] Clearly, China wants to leap ahead of the biggest aerospace power, the USA. With nearly 400,000 personnel and nearly 2,000 aircraft, of which 800 are 4th-generation plus, and nearly 150 airbases, PLAAF is significant in size. The thrust is on net-centric warfare and the need to operate in the cyber-warfare and electronic attack domains. The PLAAF commanders are conscious that it still remains a branch of the politically much more connected People's Liberation Army (PLA) which pushes its Army mindset on aviation decision making. The basic structure of the PLAAF remains in the form of departments which include elements from the Communist Party's political wing. Besides, the flying units are structured like army regiments. Many PLAAF commanders would prefer the force to be called the Chinese Air Force.

PLAAF Exposure and Strategy

China has undoubtedly made rapid progress. Some experts, however, doubt its ability to fight as a coherent force or the experience of its troops. Chinese Armed Forces have limited exposure to exercises with major world powers. They have last seen major combat engagements only in the Korean war of 1950. Maintaining such a huge military manpower and significant numbers of legacy systems, takes away a large chunk of the annual defence budget for revenue expenditure. The defence of mainland remains with the Chinese armed forces, and, in turn, PLAAF's primary mission. The east coast remains the primary area of threat and priority. Reclaiming the ROC 'Taiwan' and retaining economic and military control over the South China Sea (SCS) are major strategic goals. The Pacific remains the main area of threat from the USA and Japan. There are territorial issues with Japan also. China targets to push Forward Edge of Battle Area (FEBA) into enemy territory. It believes that greater use of SSMs makes it difficult for the enemy to defend itself. Reliance on net-centric air war backed by AWACS and FRA coupled with own secure global satellite navigation system will help dominate the air war. It's territorial interests against India are primarily in Arunachal Pradesh area and to protect China National Highway 214 (G214) which runs through Aksai Chin. PLAAF will concentrate its operations in these two areas. Chinese aircrew have had very little war-like exposure for many decades, and nor have they participated in air exercises with any major Western air force. Therefore, PLAAF now puts greater emphasis on technical knowledge and training for aircrew.

PLAN

China is known to have over 500 naval craft. With over 250,000 personnel, the People's Liberation Army Navy (PLAN) is a significant maritime force. Roughly 26,000 of these are known to be employed in the air arm. China's first aircraft carrier was the Liaoning Type 001. It was a refit of the Russian Kuznetsov-class aircraft cruiser 'Riga' and commissioned in PLAN in 2012. The Chinese used it mainly for training, experimentation, and to gain familiarity. Since 2016, it has been combat ready. Chinese maritime fighters (J-15s) have been operating onboard. Shandong (Type 002) is the first indigenous Chinese aircraft carrier. Commissioned in December 2019, it can house 44 aircraft. It has an active electronically scanned radar array (AESA) system. Like its predecessor, it uses the simpler 'short take-off but arrested recovery' (STOBAR) launch and recovery system, and therefore cannot take on the heavier J-31 aircraft. PLAN mostly uses older PLAAF aircraft like the J-10, J-15, H-6, IL-28, Ka-28, Ka-31 and KJ 200. Future carriers like the Type 003 (70,000 ton) and Type 004 (110,000 ton) would use the electromagnetic aircraft launch system (EMALS) and host heavier J-20 and J-31 class of aircraft. It is projected that China may possess five or six aircraft carriers by the mid-2030s. China could then have a significant presence even in the Indian Ocean.

China-Pakistan Aviation Ties

One of the biggest offshoots of Pakistan being China's all-weather friend has been the Chinese support for Pakistan's aerospace programs. China supplied PAF F-6 (MiG 19) in 1965 and H-5 (IL-28) in the 1970s. China helped set up the Pakistan Aeronautical Complex (PAC) at Kamra in 1973. PAF got F-7 (MiG 21) from China in the 1980s. JF-17 'Thunder' is a PAF–specific jointly developed Sino-Pakistan fighter aircraft. Nearly 120 have already been delivered and the PAF targets 300 finally. Aircraft is also being offered for export. PAF also uses Chinese ZDK-03 AWACS and the K-8 flying trainers. China has also supported Pakistan in its missile development program and has supplied air-to-air missiles. It has offered the use of the Beidou satellite navigation system to Pakistan and many other countries.

Tibet Autonomous Region

Even though the east coast remains the greater priority area, China has been developing economic and military infrastructure in Tibet Autonomous Region (TAR). The Indian border with China is entirely covered by the Chinese Western Theatre Command. The geographical area of this command is equivalent to all the other four Chinese theatre commands put together. There are eight Chinese dual-use airfields in TAR which has infrastructure for PLAAF aircraft operations. Three more are under development. Most of these airfields are at altitudes of around 3-4 kilometres, thus placing maximum take-off load restrictions and requiring much longer runways. The aviation infrastructure in Southern Theatre Command, which is immediately to the east of Myanmar, is also likely to be used by PLAAF in case of an air war across the Himalayas. There are at least five Chinese airfields from where PLAAF would have to overfly Myanmar, with or without permission. The blast protection in most TAR airfields is still under development. Similarly, ground radars are far and few in the mountains. So is the case for the IAF, both for airfields and radars in the eastern region. AWACS will be important to both sides. China is militarily not fully ready. Currently it is more of posturing.

Options for India

In 1962, despite better aircraft, the IAF was not allowed to be used by the political leadership of the day. In a possible military conflict, India would have to maintain an aggressive defence approach. The IAF which is currently at an all-time low of just 29 fighter squadrons would have to be quickly built up to the authorized strength of 42. This itself could take around two decades. In the meantime, the IAF would fight as per its Plan 'B', that is with existing assets. Air power would also have to be used extensively for inter-theatre movement of military assets. The IAF will require to support the Indian Army (IA) for inter-valley movement of troops and heavy weapons. It has no choice but to build more infrastructure in the east and also step up operational flying training in the mountains.

In view of the existing capability gap and China still pushing ahead because of much higher military spending, India needs to review its military and geo-political options. To reduce the military capability gap, India has to increase defence spending. It needs to adopt an out-of-the-box approach to military tactics and strategy, and also invest in force multipliers. China's surplus industrial capability and shrinking Western markets make India's huge and fast growing market critical for its exports. India must leverage this Chinese requirement to its advantage. India needs to strategically balance between the USA and Russia, both of which are significant to the geo-politics of the region. While India uses different approaches to engage China, it needs to have a fresh look at its own regional equations with China's neighbours.

REFERENCES

1. Carter Chapwanya, China's rise and the and the Napoleonic prophecy, modern diplomacy, March 07, 2019, https://moderndiplomacy.eu/2019/03/07/chinas-rise-and-the-napoleonic-prophecy/
2. Sriparna Pathak, The "Peace" in China's Peaceful Rise, *International Relations*, October 15, 2015, https://www.e-ir.info/2015/10/15/the-peace-in-chinas-peaceful-rise/
3. Robert L. Suettinger, The Rise and Descent of "Peaceful Rise", Hoover Institution, China Leadership Monitor, https://media.hoover.org/sites/default/files/documents/clm12_rs.pdf.
4. Deborah Welch Larson, Will China be a New Type of Great Power?, *The Chinese Journal of International Politics*, Vol. 8, Issue 4, Winter 2015, pp. 323–348, Published November 05, 2015, https://doi.org/10.1093/cjip/pov010
5. Zheng, Bijian (2005), China's Peaceful Rise to Great-Power Status. *Foreign Affairs*, Vol. 84 (5) September/October 2005, pp. 18-24. Also see Zheng, Bijian, 2005. *China's Peaceful Rise*. Washington: Brookings Institution.

6. Max Fisher and Audrey Carlsen, How China Is Challenging American Dominance in Asia, *The New York Times*, March 9, 2018, https://www.nytimes.com/interactive/2018/03/09/world/asia/china-us-asia-rivalry.html
7. Anti-Secession Law (Full text) (03/15/05), Embassy of People's Republic of China in the United States of America. http://www.china-embassy.org/eng/zt/999999999/t187406.htm
8. White Paper on Peaceful Development Road, China.org.cn, December 22, 2005 http://www.china.org.cn/english/2005/Dec/152669.htm
9. Ana Nicolaci da Costa, BBC Business Reporter, How the world is grappling with China's rising power, *BBC News*, October 26, 2018, https://www.bbc.com/news/business-45948692
10. Ibid.
11. Ibid.
12. Ibid.
13. Ibid.
14. Ibid.
15. G. John Ikenberry, Albert G. Milbank Professor of Politics and International Affairs at the Woodrow Wilson School, The Rise of China and the Future of the West: Can the Liberal System Survive? January/February 2008 issue of *Foreign Affairs*. https://wws.princeton.edu/news-and-events/news/item/ikenberry-foreign-affairs-article-rise-china-and-future-west
16. Daniel Blumenthal, Director of Asian Studies at the American Enterprise Institute, The Unpredictable Rise of China, *The Atlantic*, February 3, 2019, https://www.theatlantic.com/ideas/archive/2019/02/how-americans-misunderstand-chinas-ambitions/581869/
17. Jonathan Tepperman, China's Great Leap Backward, *Foreign Policy*, October 15, 2018, https://foreignpolicy.com/2018/10/15/chinas-great-leap-backward-xi-jinping/
18. Same as Ref. 16
19. Nick Childs and Tom Waldwyn, China's naval shipbuilding: delivering on its ambition in a big way, Military Balance Blog, The International Institute of Strategic Studies, May 01, 2018, https://www.iiss.org/blogs/military-balance/2018/05/china-naval-shipbuilding
20. Ben Westcott, China's military is going from strength to strength under Xi Jinping, *CNN*, March 5, 2019, https://edition.cnn.com/2019/03/04/asia/china-military-xi-jinping-intl/index.html
21. PTI, Beijing, At $177.6 bn, China's 2019 defence budget is over 3 times that of India, *Business Standard* March 5, 2019, https://www.business-standard.com/article/pti-stories/china-hikes-defence-budget-by-7-5-per-cent-to-usd-177-61-billion-119030500095_1.html
22. Hong Kong (CNN), China's military is going from strength to strength under Xi Jinping, March 4, 2019, https://lite.cnn.com/en/article/h_db196e0354ca52725368366a02e08a48
23. Li Jingxuan, China's limited defence spending poses no threat to any country, *China Military Online*, March 05, 2019 http://eng.chinamil.com.cn/view/2019-03/05/content_9441532.htm
24. Lalit K. Jha, Pakistan owes $10 bn debt to China for Gwadar port, other projects: Top US general, *Live Mint*, March 15, 2019, https://www.livemint.com/news/world/pak-owes-10-bn-debt-to-china-for-gwadar-port-other-projects-top-us-general-1552666074599.html

25. Scott Kennedy, Senior Adviser and Trustee Chair in Chinese Business and Economics, Made in China 2025, Centre for Strategic and International Studies (CSIS), June 1, 2015, https://www.csis.org/analysis/made-china-2025
26. Fortune Global 500, https://fortune.com/global500/2016/aviation-industry-corp-of-china/
27. Patrick Tucker, US Defence R&D Funding Falls As China's Keeps Growing, *Defence One*, February 10, 2020, https://www.defenseone.com/technology/2020/02/us-defense-rd-funding-falls-chinas-keeps-growing/163021/
28. PTI, DRDO seeks enhanced budget, says China spending more on R&D, July 11, 2018, https://economictimes.indiatimes.com/news/defence/drdo-seeks-enhanced-budget-says-china-spending-more-on-rd/articleshow/49074188.cms?utm_ source=contentofinterest&utm_medium=text&utm_campaign=cppst
29. Charlie Campbell, From Satellites to the Moon and Mars, China Is Quickly Becoming a Space Superpower, *Time*, July 17, 2019, https://time.com/5623537/china-space/
30. Andrew Jones, China prepares to launch new rockets as part of push to boost space program, Space.com, February 17, 2020, https://www.space.com/china-space-program-new-long-march-rockets.html
31. David Lee, Global BeiDou Satellite System Will Be Completed By 2020, *Pan Daily*, June 10, 2019, https://pandaily.com/global-beidou-satellite-system-will-be-completed-by-2020/
32. Frank Chen, Chinese Aero Group eyes world's largest plane, *Asia Times*, July 18, 2019, https://asiatimes.com/2019/07/chinese-aero-group-eyes-worlds-largest-plane/
33. PTI Beijing, China develops its own Mother of All Bombs, *India Today*, January 5, 2019, https://www.indiatoday.in/world/story/china-develops-its-own-mother-of-all-bombs-report-1423955-2019-01-05

2

Primacy of Aerospace Power

One Who Controls Aerospace, Controls the Planet

"To have command of the air means to cut an enemy's army and navy from their bases of operation and nullify their chances of winning the war."

—Giulio Douhet

"Air control can be established by superiority in numbers, by better employment, by better equipment, or by a combination of these factors."

—Carl Andrew Spaatz

The 21st-century aerospace technologies have profoundly altered the prosecution of war and have in fact laid the foundation for a third revolution in military affairs. Warfare in the third dimension has brought about a substantial increase in range, speed, manoeuvrability, accuracy, and lethality of platforms and weapons. The increased mobility gives aerial platforms significant advantages over surface forces in terms of responsiveness, flexibility, and versatility in the application of overwhelming destructive force. Any point on the globe can now be addressed by aerospace power within hours or minutes with little or no warning, and it allows military power to be brought to bear against any aspect of an adversary's power, be it political, military, economic, or social. Employed at the strategic, operational, and tactical levels of war, either simultaneously or separately, air and space power have become integral to military operations on land or sea. Aircraft carriers have already taken a central role in naval operations. Airpower has a great role in the air-land battle doctrine of armies. "War is thus an act of force to compel our enemy to do our will," said Carl von Clausewitz.[1] Military aerospace capabilities serve to deter aggression and prevent conflict by cautioning potential adversaries that their objectives will be denied and their aggression will be decisively defeated.[2]

U.S. Air University Defines Aerospace Power

The *United States Air University* defines aerospace power as "the ability to use a platform operating in or passing through the aerospace environment for military purposes."[3] The meaning of aerospace is elaborated as "Of, or pertaining to, earth's envelope of atmosphere and the space above it...." The perspective, range, speed, and manoeuvrability of air and space platforms produce levels of mobility, responsiveness, flexibility, and versatility unique to air and space power. The key elements are centralized control, flexibility/versatility, priority, synergy, balance, concentration, and persistence. Commanders are expected to exploit the mobility, responsiveness, flexibility, and versatility of air and space power. Air and space forces must be centrally controlled by an air commander to achieve advantageous synergies, establish effective priorities, capitalize on unique strategic and operation flexibility, ensure unity of purpose, and minimize the potential for competing objectives. Execution of aerospace missions should be decentralized to achieve effective spans of control, responsiveness, and tactical flexibility.

The ability to concentrate force anywhere and attack any facet of the enemy's power is the great strength of aerospace power. Aerospace operations must be prioritized because demands for air and space forces will invariably swamp air commanders.[4] The proper application of air and space forces, in mutually supportive air campaigns, produces effects well beyond the proportion of each mission's individual contribution to the campaign. The proper application of air and space forces in joint operations with surface forces produces effects that greatly magnifies the contribution of the surface force, in effect shape the theatre in ways that cannot be achieved by surface forces. The versatility and flexibility of aerospace power permits the simultaneous accomplishment of strategic, operational, and tactical effects through dispersed and parallel attacks in order to achieve an overwhelming effect. Air and space power should be applied persistently. Destroyed targets can be rebuilt and need to be visited through re-strikes. A primary objective of air and space operations is to gain air and space superiority, which provides the essential security and freedom of action for all friendly forces in the theatre. The enemy has to be denied access or full use of air and space. Action against a wide array of military, economic, political, and social targets permits the shaping of a conflict in ways that cannot be done by surface forces.

USAF Doctrine: Primary Roles of Aerospace Power

The U.S. Air Force (USAF) doctrine identifies four primary roles for aerospace power. 'Aerospace control', to control the combat environment, and deny to the enemy; 'force application' to apply combat power; 'force enhancement' to multiply combat effectiveness; and 'force support' to sustain forces.[5] Aerospace control includes counter-space and offensive or defensive counter-air operations. Force application involves the use of interdiction, close air support, and strategic attack missions against enemy surface components and capabilities. Force enhancement missions are those that enhance or improve the capability of combat forces, and include airlift, air refuelling, space-lift, electronic combat, special operations, surveillance, and reconnaissance. Finally, force support encompasses logistics, combat support, on-orbit support, and base operability and defence. One of the main problems for the USAF had been to define its mission as 'air and space' separately or as 'aerospace.' The main argument for making a distinction between the two mediums is that the systems and operations of air assets are inherently different from space assets. Air assets are governed by the laws of aerodynamics and tend to be more flexible and more responsive. Space assets are governed by orbital mechanics and tend to cover wider areas of the Earth and provide a more continuous presence. The argument for defining air and space as one medium 'aerospace' points to the fact that air and space platforms perform similar missions and the characteristics of air and space assets will merge as more systems that can operate in both environments become available.

Aerospace Core Capabilities

Aerospace forces need three core capabilities, highly developed and capable combat assets, mobility/transportation, and information systems. The ability of air and space power to shape the operational level can best be described by using John Boyd's OODA (Observe, Orient, Decide, Act) loop theory.[6] The OODA loop works both at tactical and strategic levels and comprises four basic operations: observation, orientation, decision, action. Space platforms in low orbits and air platforms cannot occupy a single point in the air or space for extended periods. Air platforms are sent out for one or more missions, and after each mission the air platform must return to a friendly base for refuelling and re-arming. Space platforms in low and medium orbits must often be tasked for each pass over the theatre or area of interest. The effects from each mission must be evaluated and a decision on subsequent missions must be made. The objective of air and space superiority is not to conduct air and space operations faster than the enemy, but to deny the enemy the ability to conduct any air and space operations at all. Surface forces also typically establish the objective of denying the enemy forces the ability to operate in a theatre but that denial is usually the result of other operations. Most of the important air and space assets which are essential to operations like the air operations centres, air bases, satellite ground stations, AWACS and other high-value air assets are mostly located beyond the reach of enemy surface forces. Therefore, they can only be attacked by enemy air and space forces. The ability to observe the whole theatre of operations both close in and deep into enemy territory, including radar, and infrared radiations is unique to the air forces using satellite sensors, UAVs, AWACS, Aerostats, ground-based radar, and traditional reconnaissance assets. This huge data has to be processed and disseminated in a timely manner for decision-making.

Core Missions

Air and space superiority is a necessary precursor to all successful operations. For the air force is critical to disrupt and destroy enemy command and control to prevent effective control of all armed forces. The core competencies of precision engagement and global reach are far reaching. Close air support and interdiction destroy the ability to effectively mount any offensive land or naval combat operations. Air and space power shaped the entire theatre during Operation Desert Storm. The ability of the air forces to respond quickly with massive airlifts to remote locations around the globe is unmatched. The OODA loop also signifies communications and intelligence and their relationship to air and space operations.[7] Supporting infrastructure that provides targets, target data, integrated mission orders, and the necessary logistics and maintenance are very crucial. The non-lethal applications of aerospace power include strategic reconnaissance, electronic jamming, psychological operations, transportation, and communications. They support to degrade or deny.

WW II: Aerospace Power Evolves

The primacy of air power was first established in WW II as it was used extensively in all theatres. While air forces flew their own strategic missions, they were closely integrated with land and naval forces. Strategic bombers became an important element. Tactical control of the battlefield by air, as well as the need for air defences was crucial. Achieving air superiority over the battlefields also gives vital freedom of action to surface forces. Aircraft carriers played the central role in the war at sea. The Luftwaffe, the German Air Force, was seen by Adolf Hitler as the decisive strategic weapon. For the Battle of Britain, the Royal Air Force (RAF) started on a large expansion, with many airfields being set up. They combined the newly-developed radar with communications centres to direct their fighter defences. Similarly, the Soviet Air Force was on a defensive when Operation Barbarossa began, having to confront more modern German aircraft.[8] Later, Soviet aircraft production rose sharply. The Soviets concentrated on aircraft for tactical ground attack role mainly in support of the Army.

President Franklin D. Roosevelt gave command of the U.S. Navy to an aviator, Admiral Ernest King, with a mandate for an aviation-oriented war in the Pacific.[9] Robert A. Lovett, the civilian Assistant Secretary of War for Air, said, "While I don't go so far as to claim that air power alone will win the war, I do claim the war will not be won without it."[10] George Catlett Marshall Jr., the U.S. Army Chief of Staff, though an infantryman uninterested in aviation before 1939, became a partial convert to air power and allowed the aviators more autonomy. He authorized vast spending on planes, and insisted that American forces had to have air supremacy before taking the offensive.[11] MacArthur had been badly defeated in The Philippines in 1941–42 primarily because the Japanese controlled the sky. MacArthur vowed never again.[12] His island-hopping campaign was now determined by the range of his air force, and the first task on securing an objective was to build an airfield to prepare for the next leap. The Allies won battlefield air supremacy in the Pacific in 1943 and in Europe in 1944. That meant that Allied supplies and reinforcements would get through to the battlefront, but not the enemy's. Both Germany and the USA pioneered the use of what would come to be known as Precision Guided Munitions (PGM).

Japanese naval air power proved unexpectedly powerful, sinking the American battleship fleet at Pearl Harbour in December 1941. A raid by a powerful Japanese Navy aircraft carrier force into the Indian Ocean resulted in the Battle of Ceylon and sinking of the only British carrier, HMS Hermes as well as two cruisers and other ships effectively driving the British fleet out of the Indian Ocean and paving the way for the Japanese conquest of Burma and a drive towards India. In the battle of Midway, Japan had 272 warplanes operating from four carriers; the U.S. had 348 of which 115 were land-based. In an extraordinarily close battle, the Japanese lost four main aircraft carriers, and were forced to return home.[13]

Picture Source: U.S. Department of Interior

After 1942, the USA made a massive effort to build up its aviation forces in the Pacific, and began island-hopping to push its airfields closer and closer to Tokyo. The small geographical area of Japan, large heavily populated cities, and the concentration of munitions production made strategic bombing the preferred strategy of the Americans. The Operation 'Meetinghouse' raid was on the night of March 9/10, 1945. Nearly 100,000 Japanese were killed and another million injured, most of them civilians, when more than 300 American B-29 bombers dropped 1,500 tons of firebombs on the Japanese capital that night. The inferno the bombs created reduced an area of 41 sq. km. to ash. Moreover, by some estimates, a million people were left homeless.[14] It is estimated by some to be the single most destructive bombing raid in military history. On August 6, 1945, the 'Little Boy' enriched uranium atomic bomb was dropped on the city of Hiroshima. This was followed by the detonation of the 'Fat Man' plutonium core atomic bomb on August 9 over Nagasaki, forcing the announcement of the surrender.

The 'Fat Man'.
Picture Source: Atomic Heritage Foundation

During the Battle of the Bulge in December 1944, General George Patton said the cooperation of XIX TAC Air Force was "the best example of the combined use of air and ground troops that I ever witnessed."[15] For the first time during WW II, the fundamental assumption of the air power doctrine emerged that it was just as important as ground war. The idea of combined arms operations (air, land, sea) was well established. Hereon, the USA invaded only after certain of air supremacy. Senior ground commanders now had attached air commanders.

Technology Critical to Air Power

Very early it became clear that technology was critical to air power and air supremacy. Building on their lead in ground radar, the British made the first airborne ground-scanning radar system that improved navigation, targeting and bombing at night even in bad weather. The post-war shift to jet engines resulted in enormous increases in speeds and altitudes of aircraft. High-altitude observation aircraft like the Lockheed U-2, and, later, the Lockheed SR-71 Blackbird were used for intelligence-gathering. The Soviets finally downed a U-2 piloted by Gary Powers using an SSM. Air combat also got transformed with greater use of Air-to-Air Missiles (AAM) with increased sophistication in guidance and increased range. The emphasis then shifted to highly manoeuvrable attack aircraft that could also fly 'under the radar' at very low levels. Swept-wing fighters could fly much faster. Anti G-suites increased the 'G' tolerance of pilots, and thus came highly manoeuvrable aircraft.

Indo-Pak Air Wars

The 1965 Indo-Pakistan war saw the first full-scale engagement in combat. The IAF had numerical superiority, and the PAF had a little qualitative advantage. Neither side could achieve air superiority. However, by the 1971 Indo-Pakistan war, the IAF had both the qualitative as well as numerical edge with large numbers of MiG-21s in service. After the initial pre-emptive strike, the PAF adopted a defensive stance. The IAF continued aggressively against the PAF which offered little in retaliation. The PAF failed to intervene during the Indian Navy's two raids on Pakistani naval port city of Karachi during Operation Trident and Operation Python. The PAF was unable to support the Pakistan Army during the tank battle of Longewala. The IAF's air superiority in the east destroyed the PAF. Air superiority in the east also allowed the IAF to provide unhindered support to the Indian Army for the ground offensive. It hastened the operations by providing heli-lift for troops to cross river obstacles and the famous Tangail para airdrop on 11 December 1971 when the 2nd Battalion of the Indian Army's Parachute Regiment was dropped to cut the Pakistani 93rd Brigade. The final nail in the coffin was the four-MiG-21 strike by IAF that blasted the Governor's House with 57 mm rockets on 14 December when a crucial meeting was on. The Governor of East Pakistan, Mr. A.H. Malik, was so shocked after the incident that he resigned on the spot. It was also a signal for the Pakistani surrender.

Black & White Print of a painting of the Longewala Air Strikes by Gp Capt Deb Gohain (Retd)

Air Power in Vietnam War

The large-scale use of helicopters by the US Army in Vietnam led to a new class of air-mobile troops,[16] and the introduction of 'Air Cavalry' through extensive use of the Bell UH-1 Huey helicopter. The Sikorsky CH-54 Tarhe 'Skycrane' and the Boeing Vertol CH-47 Chinook lifted heavier loads such as vehicles or artillery. The specialized AH-1 Cobra was developed from the Huey for escort and ground support. The Boeing B-52 Stratofortress was modified to unload a massive high-explosive payload on enemy troop concentrations. The Lockheed AC-130 became a significant gunship. The A-10 Thunderbolt II was the main ground attack platform. The carrier-based A-4 Skyhawks were extensively used by the U.S. Navy (USN). For the first time, the combat use of electronic computers aboard ships displayed comprehensive real-time battle pictures for better situational awareness. General Dynamics F-111 Aardvarks, pioneered several technologies, such as variable-sweep wings, first afterburning turbofan engines, and automated terrain following radar for low-level flight. North Vietnam very effectively used the anti-aircraft artillery, SA-2 guided missiles, and MiG fighters. US air strikes were for the first time supported by force multipliers like airborne radar platform the Lockheed EC-121 Warning Star, and FRA Boeing KC-135 Stratotanker, and specialised 'Wild Weasel' radar jamming aircraft. 'Smart' guided bombs were used later in the war. Air combat training school such as TOPGUN was set up to improve kill ratios. In addition, the USAF restarted programs to produce agile air superiority fighters such as the McDonnell Douglas F-15 Eagle.

Egypt Israeli Wars

On June 5, 1967, the Israel Defence Forces initiated Operation Focus, a coordinated aerial attack on Egypt.[17] Nearly 200 aircraft took off from Israel, catching the Egyptians by surprise, and attacked 18 airfields and eliminated roughly 90 per cent of the Egyptian air force on the ground. Later Israel attacked and decimated the air forces of Jordan, Syria and Iraq. By the end of the day on June 5, Israeli pilots had won full control of the skies over the Middle East. Israeli tanks and infantry could storm across the border and into the Sinai Peninsula and the Gaza Strip. On June 9, following an intense aerial bombardment, Israeli tanks and infantry advanced on the heavily fortified Golan Heights region of Syria and captured it. The Six-Day War, won mainly by air power, had momentous geopolitical consequences in the Middle East.

Egypt began a concentrated effort to build up its forces, inducting Russian MiG-21 jet fighters, SA-2, SA-3, SA-6, SA-7 anti-aircraft missiles, T-55 and T-62 tanks among many other weapons. On October 6, 1973, Egypt and Syria launched a coordinated attack against Israel on Yom

Kippur,[18] the holiest day in the Jewish calendar. Taking them by surprise, Egyptian troops swept deep into the Sinai Peninsula. The mobile SA-6 SAMs proved effective against low-flying Israeli aircraft until they were neutralized. The Israeli Air Force had no time to execute a Suppression of Enemy Air Defences (SEAD) operation due to the element of surprise. It lost 102 airplanes mostly in the initial strikes on the ground. Nearly 450 Arab aircraft were shot down. Most of these were shot in aerial engagements. Disengagement took place on October 28, 1973. Israel's victory came at the cost of heavy casualties. Air power had been intelligently and copiously used by both sides. Israeli losses were greatly reduced due to use of air power.

Falklands War

The Falklands War of 1982[19] was fought by Britain against Argentina at a distance of 12,781 kilometres. On 2 April 1982, Argentinean forces mounted amphibious landings, known as Operation Rosario, on the Falkland Islands. The operations lasted from 1 April 1982 to 20 June 1982. Sierra Leone allowed British task force ships to refuel at Freetown. VC10 transport aircraft landed at Banjul in Gambia while flying between the UK and Ascension Island. Chile gave support to the UK in the form of intelligence about early warning intelligence on Argentine air movements. The two British aircraft carriers 'Invincible' and 'Hermes' set sail. The whole task force eventually comprised 127 ships, including 62 merchant ships. The British had just 42 Harrier aircraft for air combat operations, against approximately 122 Argentinean serviceable jet fighters. The British lacked Airborne Early Warning and Control (AEW&C) aircraft. The RAF set up an airbase on Ascension Island in the mid Atlantic British overseas territory. They positioned a sizeable force of Avro Vulcan B Mk 2 bombers and aerial refuellers and some fighters to protect them. On 1 May, British operations on the Falklands opened with an attack on the airfield at Stanley. A Vulcan bomber from Ascension flew a 15,000-km round trip dropping conventional bombs across the runway at Stanley and returning to Ascension. The mission required repeated refuelling and required several Victor K2 tanker aircraft operating in concert, including a tanker for tanker refuelling. The raids prevented Argentina from stationing fast jets on the islands. Stanley airfield was too short to support fast jets. The Argentines were forced to launch their major strikes from the mainland. The effective loiter time of incoming Argentine aircraft was low. The British lost the Destroyer HMS Sheffield to fire following an Exocet missile hit from an

Argentine naval aircraft. During the six-week war, British carrier-based Harriers flew over 1,500 sorties. Twenty-one Argentine fixed-wing aircraft were destroyed in the air. An additional 18 Argentine fixed-wing aircraft were destroyed by British SAMs. Fifteen Argentine fixed-wing aircraft were destroyed on the ground and 14 were captured. Militarily, the Falklands conflict remains the largest air-naval combat operation between modern forces since the end of the Second World War. The most significant lessons were the vulnerability of surface ships to anti-ship missiles and the challenges of coordinating logistical support for a long-distance projection of power, and reconfirmation of the role of tactical air power, including the use of helicopters.

Gulf War in 1991

The collapse of the Soviet Union in 1991 brought a shift in concept from large numbers to lesser multi-role aircraft. Aircraft carriers became important in global air power projection in the absence of dedicated air bases. The PGMs allowed strikes at selected surface targets with high accuracy. The role of air power in modern warfare was dramatically demonstrated during the Gulf War in 1991.[20] Air attacks were made on Iraqi command and control centres, communications facilities, supply depots, and reinforcement forces. Air superiority over Iraq was gained before major ground combat began. The initial strikes were composed of Tomahawk cruise missiles launched from ships. Lockheed F-117 Nighthawk stealth bombers with laser-guided bombs and anti-radar missiles first destroyed the air defence network and resulted in air superiority over the country. A-10 Thunderbolt IIs attacked Iraqi armour with Gatling guns and Maverick missiles to support the advance of US ground troops. Attack helicopters fired laser-guided Hellfire missiles and TOW ((Tube-launched, Optically tracked, Wire-guided) missiles. The allied air fleet also made use of AWACS and the B-52 bombers. The allied aerial strike force was made up of over 2,250 combat aircraft, which included 1,800 US aircraft. The Iraqi force was of about 500 aircraft composed mainly of MiG-29 and Mirage F1 fighters.[21] More than 88,000 combat missions were flown by allied forces with over 88,000 tons of bombs dropped by the end of the fifth week. Air superiority had been well established resulting in low ground casualties.

Operation Safed Sagar

On 11 May 1999, the IAF was called by Indian Army to provide helicopter close air support against Pakistani intruders for its 'Operation Vijay' in Kargil. The IAF subsequently launched air operations and were called operation 'Safed Sagar.'[22] The IAF had been carrying out routine electronic intelligence (ELINT), photo and aerial reconnaissance since early May. On 21 May, a Canberra PR57 on a reconnaissance mission, was hit by a Chinese-made Anza infrared SAM. The plane returned to the nearest IAF base, Srinagar, on one engine. On 25 May, the Indian Government cleared limited use of offensive and defensive air power, restricted entirely to the Indian side of the Line of Control. The aim was to target Pakistani bunkers in the occupied hills and cut off Pak transit lines. The first strikes were launched on 26 May, and the IAF struck infiltrator bunkers with Mi-17 armed helicopters. This was followed with fighter strikes by MiG-27s and MiG-21s with MiG-29s providing air cover. On 27 May, the IAF lost a MiG-27 which suffered an engine flameout due to disturbance of air flow near the air intake on rocket firing at very high altitude. It also lost a MiG-21 to enemy ground action when the IAF pilot was trying to search out the downed MiG-27. The following day, a Mi-17 was lost when hit by three Stinger missiles while on an offensive sortie in the Tololing sector. These losses forced the IAF to reassess its strategy. It withdrew helicopters from offensive roles. On 30 May, the IAF inducted Mirage 2000 for high altitude weapon delivery. Mirage 2000s could carry out day and night attacks on Tiger Hill and Muntho Dhalo using LGBs.[23] It used the much cheaper US Pave way II LGB kits with the Israeli Litening Laser designator pods (LDP). The LDPs were also used for recce, and attack of the large Pakistani logistics camps. The attacks on Muntho Dalo and Mashkoh Nalla had a demoralizing and paralyzing affect on the Pakistani Army. The first Mirage 2000 LGB mission, was carried out by then Wg Cdr R. Nambiar (later Air Marshal), and Sqn Ldr Patnaik (now Air Marshal) and observed by the (then) Chief of the Air Staff, Air Chief Marshal A Y Tipnis, in yet another two-seater. All LGBs were delivered by two-seaters, with the rear-seat pilot doubling up as a WSO. The first night LGB mission was flown on 28 Jun achieving a direct hit, and this mission was considered instrumental in paving the way for ground forces to capture Tiger Hill. During an aggressive patrol, two MiG-29s armed with Beyond Visual Range (BVR) missiles got a missile lock on two F-16s of the Pakistan Air Force (PAF) patrolling close to the border, but did not engage them because there was no official declaration of war. By keeping the PAF threat at bay, the IAF was able to deliver numerous devastating strikes on intruder positions and supply dumps. All aircraft operated at an altitude of 10,000 metres above sea level, diving when required and pulling out well out of MANPADs range. The air strikes against the Pakistani infiltrators, supply camps and other targets yielded rich dividends. The Mirage 2000 was looked upon as the aircraft that turned the tide of the Kargil War. By the end of July 1999, India regained all its possessions lost in the Kargil region. As per Indian government reports, IAF transport aircraft flew 3,427 sorties, helicopters 2,474, and fighters 1,730 sorties, totalling 7,631.[24] The breakdown of IAF fighter missions were 48 per cent for air strikes, 30 per cent CAP and Air Escort, and 13 per cent Reconnaissance. The restrictions for not being able to cross the LoC which was very close to the area of action did place significant attack limitations on air power. Finally, air power not only greatly reduced Indian Army casualties but also turned the tide of war and hastened the end. The lessons learned in this limited war influenced India to urgently upgrade its combat fleets. The IAF also acquired and started co-developing the Sukhoi Su-30 MKI. The process of acquiring 126 Medium Multi-role Combat Aircraft (MMRCA) was also initiated in 2001.

Picture Source: Indian Air Force

Iraq War 2003 'Iraqi Freedom'

The 2003 invasion of Iraq was led by US and British forces putatively to defeat the regime of Saddam Hussein. Aerial warfare once again proved to be decisive. The air campaign began on March 19 with limited night bombing on the Iraqi capital, Baghdad. Later, intensive bombardment began. About 14,000 sorties were flown. Eight hundred Tomahawk cruise missile, each costing around a million dollars each, were fired at numerous targets in Iraq in just a period of a month.[25] Iraqi resistance was crushed. Iraqi anti-aircraft weapons were unable to open fire on high-altitude US bombers such as the B-52, or the stealth B-2 bomber and F-117A.[26] U.S. and British aircraft used radar-detecting devices and aerial reconnaissance to locate and destroy Iraqi anti-aircraft weapons. Bunker buster bombs, designed to penetrate and destroy underground bunkers, were dropped on Iraqi command and control centres. Iraqi ground forces could not seriously challenge the American ground forces because of their air supremacy. By mid-April 2003, U.S. and British forces controlled all of Iraq's major cities and oil fields.

War in Afghanistan

The war in Afghanistan began with the rise of al-Qaeda and the Taliban which rose from the ashes of Afghanistan's post-Soviet civil war in 1999. The Taliban had been providing al-Qaeda sanctuary of operations since 2001. The UNSC adopted Resolution 1267[27] which created the al-Qaeda and Taliban Sanctions Committee, which linked the two groups as terrorist entities and imposed sanctions on their funding, travel and arms shipments. The assassination of Ahmad Shah Massoud, commander of the Northern Alliance, an anti-Taliban coalition, dealt a blow to the anti-Taliban resistance and operations. Osama bin Laden started getting protection from the Taliban after the 9/11 attacks. After the 9/11 al-Qaeda attacks in the U.S.A., U.S. President George W. Bush vowed to "win the war against terrorism". It was then decided to invade Afghanistan. On 07 October 2001, U.S. military, with British support, began a bombing campaign against Taliban forces, officially launching Operation Enduring Freedom.[28] Many other allies promised to support. The U.S. air strikes were supported by over a thousand special forces troops. The Northern Alliance and other anti-Taliban forces also joined. The Taliban soon started losing ground and finally collapsed. Despite the collapse, al-Qaeda leader Osama bin Laden continued to hide in the Tora Bora caves, but he later escaped. In December 2001, Hamid Karzai was installed as the interim administration head. Operation Anaconda was the first major ground assault in early 2002, but the Pentagon's priority shifted to Saddam's Iraq. The new strategy in Afghanistan was to rebuild with humanitarian and reconstruction assistance. Meanwhile Osama bin Laden resurfaced and released a video tape claiming responsibility for 9/11. The year 2006 saw a resurgence of Afghan insurgency. However, in the renewed offensive, there were many instances of collateral civilian killings in air strikes that brought negative publicity and worldwide condemnation. A tactical directive issued by Gen. Stanley A. McChrystal on July 6, 2009 cautioned to avoid the "trap of winning tactical victories, but suffering strategic defeats, by causing civilian casualties and alienating the people".[29]

In February 2009, the new American President, Barack Obama, declared a recommitment to the war and deployed 17,000 more troops to Afghanistan, adding to the 36,000 already there. In December, Mr. Obama announced a "surge" meant to build and train an Afghan security force that would be strong enough to assume responsibility for fighting the insurgency. His plan included sending 30,000 more American troops, bringing the total number to nearly 100,000 by mid-2010.[30] The new strategy was to link success in Afghanistan to a stable Pakistan. The aim was to disrupt and dismantle safe havens in Pakistan. A fresh offensive was launched by nearly 4,000 U.S. Marines to restore government control in Helmand Province. Obama put a timeframe of July 2011 to complete operations in Afghanistan. On 01 May 2011, al-Qaeda leader Osama bin Laden was killed by U.S. forces in his hideout in the military cantonment in Abottabad through a mid-night air operation deep in Pakistan. In early 2013, an unsuccessful attempt was made to strike a deal with the Taliban. However, gradually, Afghanistan security agencies were handed control of internal security with the full backing and training by NATO allies. But U.S. war on terror continued. Meanwhile, the Islamic State started emerging as a group in Iraq and Syria. On April 13, 2017, the USAF dropped the most powerful non-nuclear bomb, also referred to as the 'the mother of all bombs,'[31] on an Islamic State cave complex in eastern Nangarhar Province. After his election, President Trump initially signalled a pullout, but later supported a prolonged Afghan war. He invited India to play a greater role in rebuilding Afghanistan. Meanwhile, the Taliban carried out some bold attacks amid U.S. escalation.

The new rules of engagement had made it easier for U.S. forces to carry out strikes against the Taliban. Also, resources had shifted from the fight against the Islamic State (IS) in Syria and Iraq to Afghanistan. At the same time, the USA had launched a five-year plan to massively expand and overhaul the Afghan Air Force, including providing it with 159 Black Hawk helicopters. John W. Nicholson, the top U.S. General in Afghanistan, pledged that a "tidal wave of air power" will be unleashed to push the Taliban to the negotiating table, and perhaps bring an end to America's longest war—which has dragged on for 17 years.[32]

Collateral civilian damage had been an issue. Ever since the negative media coverage of collateral damage of loss of innocent civilian lives in the Vietnam war, weapons have been developed with higher accuracy. Heavy bombing against the Taliban and IS saw many Afghan civilians killed and injured from the air, though most civilian casualties were caused by anti-government groups like the Taliban and IS. There are some who believe that the entire strategy of pounding the Taliban militarily was misguided. A BBC study found that Taliban fighters continued to be openly active in 70 per cent of Afghanistan.[33] Barnett Rubin, who served as senior adviser to the Special Representative for Afghanistan and Pakistan at the US Department of State from 2009-2013, said the air campaign was having "no strategic effect."[34]

Air Power: Capital Budgets

Air power continues to dominate most conflicts around the world. The investments in air power even by land and maritime forces are also significantly increasing. The biggest share of capital procurement budgets are also for air power assets. U.S. DoD budget proposals for 2021 at $ 705.4 billion[35] have major funding for Nuclear Modernisation ($28.9 billion) that includes B-21 Bomber. Missile Defence ($20.3 billion), the new Space Program $18 billion, while Cyberspace protection is budgeted at $9.8 billion, which includes AI and Cloud. The Air Domain $56.9 billion includes $11.4 billion for 79 F-35 Joint Strike Fighters. The Marine Domain $32 billion includes $3.0 billion for an aircraft carrier. The land domain is budgeted for $13 billion. The Department's largest RDT&E budget, $106.6 billion, is the largest ever focused on the development of crucial emerging technologies, including $3.2 billion for Hypersonics. The services readiness (revenue) budgets are Army $30.9 billion, Navy and Marine Corps $47.5 billion, Air Force $37.1 billion, and $69 billion for Overseas Contingency Operations.

China's biggest portion of the defence budget goes into the development program of aircraft carriers, fighter aircraft and other aerial platforms. The IAF gets the highest percentage of the capital budget at 57 per cent of its total allocation for Financial Year (FY) 2019-20. Aerial platforms will affect the outcome in war and peace, said Air Chief Marshal Arup Raha, Chief of the Air Staff, IAF.[36] He stressed the need for procuring aerial strategic assets on a continuous and long-term basis.

IAF Doctrinal Shift: Greater Offensive Role

The IAF has also seen a doctrinal shift over the years. It will now take on a more offensive role against both adversaries. Introduction of combat enablers like AWACS, flight refuellers and strategic transport aircraft has made a difference. In the recent, so called mother-of-all-exercises, 'Gagan Shakti' the IAF tested out a possible two-front war and included inter-theatre movements of assets among practically all other domains of modern air power. Regional power projection and ability to take on China are the clear new doctrinal focus areas. The IAF's 2012 doctrine[37] clearly brings out the intent to dominate conflict and a greater role of air power in the full spectrum of national security and diplomacy. The IAF sees that deterrence and control of the air are linked. Its control of the air may not be absolute, but it hopes to possess sufficient control in order to prosecute the campaign. IAF transformation is being driven from just being platform-based to being capability-based. Effects based, network centric operations are the new normal. The advantage of air power is the ability to exploit swing-role capabilities. Modern platforms are critical. As Air Chief Marshal N.A.K. Browne said, "Our aim is to preserve and maintain, upgrade and improve, and replace and acquire. At the end of this process, IAF force structure will be modern and potent with new and upgraded fleets only".[38] This action plan is underway, albeit a little slowly, because of fund crunch. Induction of high technology assets is on and existing fleets are going through mid-life upgrades.

Any modern air force must have at least 40 per cent combat fleet comprising high tech all weather platforms. Another 40 per cent upgraded fleets still having 20 years of life ahead and the remaining 20 per cent under changeover but still giving strength through numbers. The heavy lift capability for both transport and helicopter fleets is crucial for inter/ intra theatre movements in war and sustaining troops and relief operations during peace time. A game-changer shift took place on 26 February 2019, when India decided to use offensive air power by carrying out deep strikes against targets in Pakistan even beyond Pakistan-occupied Kashmir (PoK). The Pakistani bogie of nuclear retaliation and over-hang even against a conventional strike was demolished.

Air Power Against Hybrid Threats

Air power has now to contend with 'hybrid threats' that take advantage of modern technologies for espionage, sabotage, data collection, attacks on people or infrastructure, or the achievement of media effects. Hybrid warfare exploits the vulnerabilities of the system. The air domain is being increasingly exploited to perpetrate illicit and terrorist attacks. Commercial-off-the-shelf (COTS) technologies have greatly improved the aerial warfare capabilities of non-state actors. For long it was thought that air power could only support land forces to counter hybrid threats. However, it is now clear that air power can act independently against hybrid threats. Air policing, air defence, and intervention against civilian aircraft and against other hybrid threats are evolving. Air commanders have the means to act against terrorists, and other illegal, hazardous and dangerous acts in the air domain. National resilience and the nation's will to prevent and deny threats represent a strong deterrent in hybrid warfare. Transforming the traditional concept of AD with an appropriate level of flexibility to manage the new aerial hybrid threats is evolving. Satellite and aerial surveillance, air traffic management, cyberspace monitoring, low RCS detection, multi-sensor fusion and tracking, and finally kinetic and non-kinetic engagement using laser beams or directed energy weapons are some of the means.

Technology Driving Air Power

Today, technologies are offering enhanced capabilities that are driving operational employment and tactics. AI, cyber, hypersonic flight and weapons, DEW, smart structures, and hybrid systems will dictate the future. Demand for streaming high-quality data requires bandwidth, which involves innovating sensor/processing systems. Thermally efficient, high-performance mission computer systems and network-centric payload processing units enable onboard data fusion prior to sending to digital links. Next-generation avionics would be smaller, more efficient and capable of operating under extreme conditions. Gallium Nitride (GaN) is a semiconductor material that is more efficient, easier to cool, and improves reliability for radars. Any system must be designed to maintain a competitive advantage in an austere budget environment. The Passive Aero-elastic Tailored[39] (PAT), a uniquely designed composite wing, will be lighter, more structurally efficient and have flexibility compared to conventional wings. This wing will maximize structural efficiency, reduce weight and conserve fuel. Hypersonic cruise, fuel cell technologies, hybrid sensors, improved human-machine interface using data analytics and bio-mimicry, combination of materials, apertures and radio frequencies that ensure survival in enemy territory are under development. Things will be built faster, better and more affordably, using 3D printing yet ensuring quality and safety standards. Additive 3D manufacture creates a world with spare parts on demand, faster maintenance and repairs, more effective electronics, and customized weapons. The development of a hypersonic aircraft would forever change the ability to respond to conflicts. Nano-materials will control sizes, shapes and compositions, and significantly reduce weight yet create stronger structures for air and spacecraft, yet drive down costs.

Sixth Generation Air Platforms

The Americans are already working on sixth generation aircraft which would have enhanced capabilities in reach, persistence, survivability, net-centricity, sustained awareness, human-system integration and weapon effects. Boeing's Phantom Works and Lockheed Martin's Skunk Works are already having work-in-progress. The projects are currently limited to concept images. The USAF's current approach to the sixth-generation fighter is the "Digital Century Series".[40] Instead of maturing technologies over time to create an exquisite fighter, the Air Force's goal would be to quickly build the best fighter that industry can muster over the next few years, integrating whatever emerging technology exists. The service would select, put a small number of aircraft under contract and then restart another round of competition which would revise their fighter designs and explore newer leaps in technology. The result would be a networked family of fighters. One jet might be optimized around an airborne laser, another fighter might prioritize state-of-the-art sensors and include AI. One might be an unmanned weapons truck. Instead of trying to hone requirements to meet an unknown threat 25 years into the future imagine "every four or five years there was the F-200, F-201, F-202".

Picture Source: NextBigFuture.com

The USN has a program called 'Next Generation Air Dominance',[41] to complement the smaller Lockheed F-35 and replace the Boeing F/A-18E/F Super Hornet. The current rationale is to look at both new airframes as well as adaptations of the best of technologies that are available. Industry developers are already building prototypes of 6th-Gen fighters with newly designed, stealthier airframes. Looking at applications of AI, there will be miniaturized long-range sensors, targeting technology and drones operating with ever-increasing levels of autonomy.

The future system will have to counter adversaries equipped with next generation advanced electronic attack, sophisticated integrated air defence systems, passive detection, integrated self-protection, directed energy weapons and cyber-attack capabilities. Aircraft must be able to operate in the anti-access/area-denial environment that will exist in the 2030–50 timeframe. The USAF's budget request for fiscal 2020 shows the service had asked for $1 billion for next generation air-dominance projects.[42] New proposals call for greater speed, range, stealth and self-healing structures; developments that will require new breakthroughs in propulsion, materials, power generation and weapon technology. Self-healing structures in particular would pose a significant advantage over modern-day aircraft, remaining airborne despite taking heavy fire. Avionics are supposed to withstand next generation electronic attacks and cyber-attacks and have passive detection, and integrated self-protection. The tailless flying wing, "cranked kite" design concept currently appears the way forward for future fighter aircraft. New generation of engines will allow ultra-high altitude super-cruise. Sixth-generation fighters are expected to use advanced engines such as 'Adaptive Versatile Engine' technology to allow longer ranges and higher performance, where the ratios of bypass and compression airflow can be made variable to improve efficiency.

Future weaponry would utilize scramjets for the production of faster missiles at speeds at which a missile could not be stopped by conventional air defence technology. Continued experiments with DEW and lasers, used for defensive as well as offensive measures, are also likely to shape precisely what sixth generation fighters are equipped with.

The USAF inducted its first fifth generation aircraft, the F-22, in 2005. In July 2015, the US Marines declared its first squadron of F-35B fighters operational. The USAF first squadron of F-35As became operational in August 2016. The USA plans to buy 2,663 F-35s, which will provide the bulk of the crewed tactical airpower of the USAF, Navy and Marine Corps in the coming decades. Deliveries of the F-35 for the U.S. military are scheduled until 2037 with a projected service life up to 2070. The Russian fifth generation aircraft PAK FA first flew in 2010. The Russian Air Force has just formed the first squadron with its variant Su-57. The Russian sixth generation fighter Mikoyan MiG-41 is currently under development. If selected by the Russian Air Force, the first production MiG-41 could enter service around 2027.

The Chinese fifth generation fighter, Chengdu J-20, entered PLAAF service in September 2017. A 2015 Rand Corporation report noted the J-20's combination of stealth and long range could hold USN surface assets at risk, and that a long-range maritime strike capability may be a cause for greater concern than a short-range air-superiority fighter. After the deployment announcement, several analysts noted that the J-20 will let China leverage a significant edge over India, Japan and South Korea, which struggle to design and produce their own fifth-generation fighters program by schedule. Some of the technologies evolved in J-20 are being offered to Pakistan in the Block III JF-17 fighter. The Shenyang FC-31 is still undergoing flight testing, and PLAAF or PLAN have still to commit for the program. Meanwhile, China has reportedly begun development on a 6th generation fighter, and has announced important breakthroughs in designing and developing several key components, including a next-generation engine. China plans to field it in the 2025–2030 time frame. France has reportedly abandoned any attempt to develop an indigenous fifth-generation fighter and has moved resources directly to development of a sixth-generation fighter aircraft. On March 22, 2016, Japan conducted the first flight of the Mitsubishi X-2 Shinshin test-bed aircraft for its sixth generation aircraft project.

Indian FGFA

India has walked out of the Russian Fifth Generation Fighter Aircraft (FGFA) project over cost, work share and core technology transfer issues. In any case, India has already committed to the 272 Su-30 MKI. That currently amounts to nearly 45 per cent of the IAF. Putting any further eggs in the Russian basket has its own risks and dynamics. The Advanced Medium Combat Aircraft (AMCA) is still in work-in-long-term progress. The core technologies have still to be evolved or bought. India is still at the LCA Mk 1 stage. A big step in-between is the LCA Mk 2, which may take over 6-7 years for its first flight. As per DRDO estimates, which invariably are tall orders the AMCA first flight will be around 2028 and that would mean induction around 2035. That would mean nearly 18 years after the Chinese J-20. India can ill afford that long wait. The Americans are currently not talking with India on the F-35. They believe that India must first fit into the American fighter aviation eco-system, tacitly implying that India must first choose between the F-21 (India-specific variant of the F-16) or the F-18 Super Hornet, or the recently on offer F-15EX. Buying a foreign fifth generation aircraft could further delay the AMCA. The best option then would be to acquire critical technologies during the new fighter acquisition even if it means paying for some. AMCA radar and EW suites can be developed through the joint venture route. By 2025, India should review the progress of the AMCA and maybe decide to buy two squadrons of the Russian or American aircraft.

A model of the AMCA. Displayed at Aero India 2013.
(Pic by Aeronautical Development Agency via Wikipedia)

Conclusion

The primacy of air power continues to be a decisive factor in shaping the outcome of future conflicts. Air forces have developed into a major 'Component of National Power', which can be applied quickly and decisively. Air superiority is the single most important factor in deciding the outcome of a modern conventional war. Military operations on land, sea, or in the air are extremely difficult, if not impossible, for the side that does not control the sky. Many decades ago, Field Marshal Bernard L. Montgomery had said, "If we lose the war in the air, we lose the war and we lose it quickly".[43] It still holds true, and magnified many times more. Gaining air superiority is not an end in itself. It is a means to an end: to damage, destroy, or otherwise affect an enemy's centres of gravity, whatever they may be. Air superiority must be won, and to maintain it requires constant investment and training. Although air superiority will not by itself bring victory, it is almost impossible to achieve success without it.

REFERENCES

1. Tom Gregg, The Philosopher of War, A Short Guide to *On War* by Carl von Clausewitz, https://medium.com/history-of-yesterday/the-philosopher-of-war-143990ad37bb
2. Major Gregory P. Cook, USAF, Waging Peace: The Non-Lethal Application of Aerospace Power, Fas.org, May 22, 1995, https://fas.org/spp/eprint/cook.htm
3. Basic Aerospace Doctrine of the United States Air Force, Air Force Manual 1-1, Volume 2 Essay 1, p. 71, March 1992, https://books.google.co.in/books?id=lU8b5GnLCrwC &pg=PA71&lpg=PA71&dq=United+States+Air+University+defines+aerospace+ power+ as+ %22the+ ability+t o+use+a+platform+o perating+in+or+passing+ through +the+aerospace+environment+for+military+purposes.%22&source= bl&ots=7TS3-iaM2r&sig=ACfU3U2EAaUfvh51At03kc6dt8MAbgH0jw&hl=en&sa= X&ved= 2ahUK Ewiowfr AnI3oAh UIXis KHT2VBNw Q6AEw AHoECAMQA Q#v=onepage &q=United%20States%20Air%20University%20defines%20aerospace %20power%20as%20%22the%20ability%20to%20use%20a%20platform%20operat ing%20in%20or%20passing%20through%20the%20aerospace%20environment%20 for%20military%20purposes.%22&f=false
4. Alec M. Robinson, Major, USAF A Research Report, AIR COMMAND AND STAFF COLLEGE AIR UNIVERSITY DISTINGUISHING SPACE POWER FROM AIR POWER: IMPLICATIONS FOR THE SPACE FORCE DEBATE, Maxwell Air Force Base, Alabama, April 1998, https://fas.org/spp/eprint/98-239.pdf
5. Basic Aerospace Doctrine of the United States Air Force, Air Force Manual 1-1, Volume 1 Section C, p. 6, March 1992, https://apps.dtic.mil/dtic/tr/fulltext/u2/ a338559. pdf
6. M.C. Baner, Defining Aerospace Power - Air University, Air and Space Operations Figure 1, https://www.airuniversity.af.edu/Portals/10/ASPJ/journals/Chronicles/ baner.pdf
7. Ibid.
8. Alan Taylor, World War II: Operation Barbarossa, *The Atlantic*, July 24, 2011, https:// www.theatlantic.com/photo/2011/07/world-war-ii-operation-barbarossa/100112/
9. Air Warfare of World War II, WikiZero, https://www.wikizero.com/en/Air_warfare_ of_World_War_II
10. Edited by Jacob Neufeld, A CENTURY OF AIR POWER LEADERSHIP Past, Present, and Future, Proceedings of a Symposium, p. 72, October 29–31, 2003, https:// media.defense.gov/2011/Feb/17/2001330161/-1/-1/0/AFD-110217-034.pdf
11. World War in the Air - Conservapedia, Air Power and U.S., https://www.conserva pedia.com/World_War_II_in_the_Air
12. Ibid.
13. Michael Ray, Battle of Midway, World War II, *Encyclopaedia Britannica*, https:// www.britannica.com/event/Battle-of-Midway
14. Brad Lendon and Emiko Jozuka, CNN, History's deadliest air raid happened in Tokyo during World War II and you've probably never heard of it, March 8, 2020, https:// edition.cnn.com/2020/03/07/asia/japan-tokyo-fire-raids-operation-meetinghouse- intl-hnk/index.html

15. Congressional Record: Proceedings and Debates, Appendix A 7099, September 25, 1962, https://books.google.co.in/books?id=aRWvl8_a8dUC&pg=SL1-PA7099&lpg=SL1-PA7099&dq=General+George+Patton+said+the+cooperation+of+ XIX+ TAC+ Air+ Force &source= bl&ots= JM2_ P2G3JT&sig= ACfU3U3v8CT_ Ti3EAMTws YIFXk 3p96hd8Q&hl=en&sa=X&ved=2ahUKEwifmLGftI3oAhVzH7cAHUNHC_wQ6AEwBnoECAoQAQ#v=onepage&q=General%20George%20Patton%20said%20the %20co opera tion%2 0of%20XIX%20TAC%20Air%20Force&f=false
16. Editors of *Encyclopaedia Britannica*, Air Cavalry, https://www.britannica.com/topic/air-cavalry
17. History.com Editors, Six-Day War, History.com, May 11, 2018, https://www.history.com/topics/middle-east/six-day-war
18. AJImpact, The October Arab-Israeli War of 1973: What happened? Aljazeera.com October 8, 2018, https://www.aljazeera.com/indepth/features/2017/10/arab-israeli-war-of-1973-what-happened-171005105247349.html
19. Katelyn Tietzen, Clemson University Tiger Prints, After Thirty Years: The Falklands War of 1982, 2014 May, https://tigerprints.clemson.edu/cgi/viewcontent.cgi? article=2922&context=all_theses
20. Diane T. Putney, Air power Advantage, Planning the Gulf war 1989-1991 campaign, Air Force History and Museum Program, USAF, 2004, Execution and Key Aspects, p. 339 onwards, https://media.defense.gov/2010/May/25/2001330265/-1/-1/0/AFD-100525-065.pdf
21. Google Books, Aviation, Aerial Warfare, p. 110, https://books.google.co.in/books?id=-aj9pekQcYMC&pg=PA110&lpg=PA110&dq=The+allied+aerial+strike+force+ was+ made+up +of+over+2,250+combat+aircraft,+which+included+1,800 +US +aircraft.&source=bl&ots=s0p4zza41p&sig=ACfU3U3Jmi8dJKL-ACw7sRewLm3L_83vSQ&hl=en&sa= X&ved=2 ahUKEwj_s5yojY_oAhXM6nMBHfMeBiwQ6AEwD3oECAkQA Q#v=onepage&q=The%20allied%20aerial%20strike%20force%20was%20 made%20up%20of%20over%202%2C250%20c ombat%20aircraft%2C%20which%20 included%201%2C800%20US%20aircraft.&f=false
22. Air Vice Marshal Satish K. Jain, Operation Safed Sagar, *Indian Defence Review*, Vol. 16.3 July-Sep 2001, http://www.indiandefencereview.com/spotlights/operation-safed-sagar/
23. Benjamin Lambeth, Airpower at 18,000′: The Indian Air Force in the Kargil War, Carnegie Endowment for International Peace, September 20, 2012, https://carnegieendowment.org/2012/09/20/airpower-at-18-000-indian-air-force-in-kargil-war-pub-49421
24. Ibid.
25. Martin Stenzano, What is a Tomahawk Missile? History Stories, History.com, August 30, 2018, https://www.history.com/news/what-is-a-tomahawk-missile
26. Google Books, Aviation, Aerial Warfare, p. 111, https://books.google.co.in/books?id=-aj9pekQcYMC&pg=PA111&dq=Iraqi+anti-aircraft+weapons+were+unable + to+open+fire+on+high-altitude+US+bombers+such+as+the+B-52,+or+ the+stealth+B-2+bomber+and+F-117A&hl=en&sa=X&ved=0ahUKEwi4k_ 6VoI_oAhV87nMBHWHuCn8Q6AEIKTAA#v=onepage&q=Iraqi%20anti-aircraft%20 weapons%20were%20unable%20to%20open%20fire%20on%20high-altitude%20US%20bombers%20such%20as%20the%20B-52%2C%20or%20the%20stealth%20B-2%20bomber%20and%20F-117A&f=false
27. United Nations Security Council, Sanctions, ISIL (Da'esh) & Al-Qaida Sanctions Committee, https://www.un.org/securitycouncil/sanctions/1267

28. CNN Editorial Research, Operation Enduring Freedom, Fast Facts, December 24, 2019, https://edition.cnn.com/2013/10/28/world/operation-enduring-freedom-fast-facts/index.html
29. NATO/ISAF Tactical Directive, July 6, 2009, https://www.nato.int/isaf/docu/official_texts/Tactical_Directive_090706.pdf
30. Rod Nordland and David Zucchino, America's Longest War, *The New York Times*, February 29, 2020, https://www.nytimes.com/2020/02/29/world/asia/afghanistan-war-photos-pictures.html
31. Helene Cooper and Mujib Mashal, U.S. Drops 'Mother of All Bombs' on ISIS Caves in Afghanistan, *The New York Times*, April 13, 2017, https://www.nytimes.com/2017/04/13/world/asia/moab-mother-of-all-bombs-afghanistan.html
32. Kevin Ponniah, BBC News, Counting the cost of Trump's air war in Afghanistan, June 7, 2018, https://www.bbc.com/news/world-asia-44282098
33. Ibid.
34. Ibid.
35. U.S. DOD Releases Fiscal Year 2021 Budget Proposal, February 10, 2021, https://www.defense.gov/Newsroom/Releases/Release/Article/2079489/dod-releases-fiscal-year-2021-budget-proposal/
36. Arup Raha, Air Force Chief, Aerial platforms will affect outcome in war and peace:, *Economic Times*, Last Updated: July 13, 2018, https://economictimes.indiatimes.com/news/defence/aerial-platforms-will-affect-outcome-in-war-and-peace-arup-raha-air-force-chief/articleshow/49028749.cms?from=mdr
37. Doctrine Team, Basic Doctrine of Indian Air Force 2012, Published by Air HQ Indian Air Force, https://www.scribd.com/doc/109721067/Basic-Doctrine-of-Indian-Air-Force-2012-PDF
38. Keynote Address by Air Chief Marshal N.A.K. Browne, Air Power and the IAF's Strategic Transformation at Manohar Parrikar Institute for Defence Studies and Analysis, September 23, 2013, https://idsa.in/keyspeeches/AirPowerandtheIndian AirForce
39. NASA Study, Experimental Wing Proving New Design Methods, September 26, 2018, https://www.nasa.gov/centers/armstrong/features/pat_wing.html
40. Valerie Insinna, The US Air Force's radical plan for a future fighter could field a jet in 5 years, *Defence News*, September 16, 2019, https://www.defensenews.com/digital-show-dailies/2019/09/16/the-us-air-forces-radical-plan-for-a-future-fighter-could-field-a-jet-in-5-years/
41. Kris Osborn, The U.S. Navy Is Demanding A Stealth Fighter Even Better Than the F-35, *The National Interest*, January 10, 2020, https://nationalinterest.org/blog/buzz/us-navy-demanding-stealth-figher-even-better-f-35-112311
42. Rachel S. Cohen, Brian W. Everstine, and John A. Tirpak, What's in the 2020 Budget Plan—and What's Not, *Air Force Magazine*, March 20, 2019, https://www.airforcemag.com/article/whats-in-the-2020-budget-plan-and-whats-not/
43. Bernard Law Montgomery Quotations, https://www.quotetab.com/quotes/by-bernard-law-montgomery

3

China's Aggressive Aerospace Push

"We must have some of our people learn aviation, for we can use them whenever we are in need".

—Dr. Sun Yat-Sen (1909)

As China's world influence expands, so does its military. Chinese aerospace developments have been making headlines for over a decade. The "Made in China 2025" industrial plan, which targets 10 high-value industrial sectors for global dominance, demonstrates that the Chinese government is doing more than merely "breaking the rules;" it is seeking to set new terms for international economic competition.[1] China has singled out the most successful American industrial product in its markets for competition, that is air and spacecraft, and the related technologies required to mass-produce them, represent perhaps the highest end of the value chain in exportable goods in the global economy in terms of value and scale required for production.[2]

China's Period of Strategic Opportunity

China's leaders have benefited from what they view as a "period of strategic opportunity" during the initial two decades of the 21st century to develop domestically and expand China's "comprehensive national power".[3] Over the coming decades, they are focused on realizing a powerful and prosperous China that is equipped with a "world-class" military, securing China's status as a great power to emerge as the preeminent power in the Indo-Pacific region. China's leaders seek to align civil and defence technology development to achieve greater efficiency, innovation, and growth. In recent years, China's leaders elevated this initiative, known as Civil Military Integration (CMI), to a national strategy that incentivizes the civilian sector to enter the defence market. The national CMI strategy focuses on hardware modernization, education, personnel, investment, infrastructure, and logistics. China's leaders are leveraging China's growing economic, diplomatic, and military influence to establish regional pre-eminence and expand the country's international influence. China's advancement of projects such as the "One Belt, One Road" Initiative (OBOR) will probably drive military overseas-basing through a perceived need to provide security for OBOR projects.[4] China's leaders are committed to developing military power commensurate with that of a great power. Chinese national interests overseas include a growing emphasis on the importance of the maritime and information domains, offensive air operations, long-distance mobility operations, and space and cyber operations.[5] PLA modernization includes command and force structure reforms to improve operational flexibility and readiness for future deployments. Its military modernization program has become more focused on investments and infrastructure to support a range of missions beyond China's periphery, including power projection, sea lane security, counter-piracy, peacekeeping, humanitarian assistance/disaster relief, and non-combatant evacuation operations.[6]

Goals of PLA Airpower

The largest aviation forces in the region, the PLAAF and PLAN Aviation, continue to work toward achieving long-range power projection capability. The PLAAF continues to modernize with the delivery of indigenous manned aircraft and a wide range of UAVs. China has begun test firing the Russia-produced S-400 long-range SAM system, and is developing the HQ-19, which will likely have a ballistic missile defence capability.[7] PLAAF and PLAN Aviation are the largest aviation forces in the region and the third largest in the world, with more than 2,700 total aircraft (not including trainer variants or UAVs) and approximately 2,000 combat aircraft (including fighters, strategic bombers, tactical bombers, multi-mission tactical, and attack aircraft).[8]

Focus on Air Superiority

China has made rapid relative improvements in air superiority in the Asian region, and modernized its airpower with more than 700 fourth generation fighters. The Chinese J-20 stealth fighter is more advanced than any other air platform currently deployed in the Asia-Pacific region. They are also testing the smaller stealth fighter, the FC-31, which is targeted to rival the technological capability of the U.S.'s F-35. China's J-10 and J-11 fighter jet aircraft could be roughly equivalent in capability to an upgraded U.S. F-15. Russian-built Su-27s and Su-30s, and the recently inducted Su-35, offer significant range. These aircraft along with J-10A/B/C will allow China the ability to conduct air superiority missions in the Taiwan Strait, and over the Himalayas. The Chinese today have very long range indigenous air-to-air missiles.

Second Priority: Strike

China's second focus area is Interdiction, and Maritime strikes, to be conducted in conjunction with PLARF, short-range ballistic missiles (SRBM) and land-attack cruise missiles (LACM) units for coordinated air and missile strikes against high value targets. A typical air campaign will start with cyber network attacks, electronic deception, electronic interference, and firepower destruction. The targets will be air bases, AD command, control centres, and SAM sites. Manned aircraft will follow in subsequent waves. Suppression of Enemy AD (SEAD) will be a focus area. Anti Radiation Missiles (ARM) will be used. The latest bomber H-6N, a heavily redesigned version capable of aerial refuelling and carrying Air Launched Cruise Missiles (ALCM) will be used. JH-7/As and J-16s would be important for the maritime strike role.

Third Priority: AEW&C and ISR

The third focus area is AEW&C, Intelligence Surveillance and Reconnaissance (ISR) and Electronic Warfare (EW). PLA medium-large AEW&C aircraft such as KJ-500, KJ-200, and KJ-2000 would seek to maintain at least two persistent 24-hour orbits within Chinese airspace for monitoring air and maritime activity and relaying information to PLA aircraft, ships, air defences, and building overall situational awareness. Two or more ELINT and Signal Intelligence (SIGINT) aircraft would also operate at standoff range within defended Chinese airspace alongside two or more Y-8G/Y-9G standoff ECM aircraft.

Fourth Priority: Game Changing Technologies

"We had an environment where we could do whatever we wanted in the air, and what the Chinese have done is to say you no longer can," said Douglas Barrie, senior fellow for military aerospace at the International Institute for Strategic Studies.[9] "Some of China's biggest strides are coming in air-to-air missiles, the weapons that for one or two million dollars can destroy a US$ 150 million aircraft. China's new aircraft, combined with the latest air-to-air, cruise, anti-ship and Russian S-400 air-defence systems (considered the world's best) "have made the ability of the USA to operate in contested areas very high risk", said Tim Heath, a senior international defence researcher at Rand.[10] Cyber-warfare, AI, and hypersonic aerial systems are other areas.

Military Aviation Industry: Counterfeit Approach

China's military-aviation industry is in the midst of a transformation that appears to be resulting, finally, in significant improvements in military-aviation production capabilities, as China finally emerges out of decades of licence-production of military platforms. To achieve this, China has often skipped expensive and time-consuming R&D, and chosen stealing or reverse engineering military technologies. In addition, it is most visible in Chinese aircraft and sub-systems designs. Having acquired technologies from the closed Israeli aircraft project 'Lavi', the J-10 has a striking visual resemblance with the F-16.[11] Of course, the current updated version of the J-10 has an advanced fire control radar array, and increased use of composite materials to reduce weight, and a number of other domestically developed updates. When the Soviet Union collapsed, China seized the opportunity to secure the production line for the Sukhoi Su-27. China quickly set about producing their own Su-27s, and then improved upon the design to develop what would become the J-11.[12] When the Soviets refused to part with their carrier-based Su-33 design, China purchased a Su-33 prototype aircraft from Ukraine, dubbed the T-10K-3, and quickly set about reverse engineering it to make the J-15. America claimed that China's Caihong-4 (CH-4) armed drone was based on stolen plans for the General Atomics MQ-9 Reaper, but later they made their own better variant (CH-5). China mixed and matched the F-16 design with MiG-21, to create FC-1 (JF-17 in Pakistan), with the F-16's nose and tail joined by a distinctly MiG-21 wing design.[13] The J-20, China's first fifth generation fighter, has copied technologies and concepts from the American F-22 Raptor. The Lockheed Martin design was stolen by a Chinese national named Su Bin.

Similarly, the F-35 Joint Strike Fighter design was also compromised leading to China's J-31 program, at least in external appearance. Although some airframes and engines are old, by Western standards, many aviation firms are also beginning to produce military systems that are comparable to aircraft in service with the world's advanced militaries. Like the U.S.A. China today deploys aircraft with a broad range of capabilities. They have, of late, invested heavily in weapons technologies.

Fifth Generation Fighter

China has an array of major military aircraft projects. The Chengdu J-20, also known as Mighty Dragon, is a single-seat, twin-engine, all-weather, stealth, fifth generation fighter designed as an air superiority fighter with precision strike capability. The J-20 made its maiden flight on 11 January 2011, and the first PLAAF combat unit was formed in February 2018.[14] It is the world's third operational fifth-generation stealth fighter aircraft after F-22 and F-35, and was the first in Asia. The Shenyang J-31 or Gyrfalcon is a twin-engine mid-size fifth-generation fighter currently under development.[15] It has also been referred to as the 'F-60' or 'J-21' in some media reports. On 31 October 2012, prototype No. 31001 conducted its maiden flight. China became the second nation to have two stealth fighter designs in field-testing at the same time. The J-31 incorporates certain stealth characteristics such as forward swept intake ramps and diverter-less supersonic inlet bumps and a two-piece canopy. The J-31 can carry 8,000 kg of payload, with four munitions totalling 2,000 kg internally, and 6,000 kg carried on six external hard-points.

J-20 Stealth Fighter.
Picture Source: South China Morning Post

H-20 Stealth Bomber

The Xian H-20, sometimes referred to as Xian H-X, is a subsonic stealth strategic bomber that is to enter service around 2025. It is expected to carry nuclear weapons. It is meant to strike targets beyond the second island chain (up to Guam) while carrying a payload of at least 10 tons. The bomber is expected to have a maximum un-refuelled range of around 8,000 km and maximum pay load anywhere between 10 to 23 tons, says aviation analyst Sebastian Roblin in the *National Interest* magazine.[16] The H-20 could use a low-probability-of-intercept AESA radar, and fuse that information to a firing platform hundreds or even thousands of kilometres away. The H-20 could also be used for electronic warfare or to deploy specialized directed energy. Aviation researcher Fu Qianshao states that China's long-range bomber should have a range of at least 12,000 kilometres and 20 tons of payload capacity.[17] The Xi'an Y-20 is a large military transport aircraft (payload 66 tonnes) which was officially launched in 2006. Within the Chinese aviation industry, the aircraft is more commonly known by its nickname 'Chubby Girl' because of its wide fuselage. In June 2016, *Jane's* magazine reported that up to 1,000 Y-20s are being requested by the Chinese military. On 6 July 2016, the first series production Y-20 (serial number 11051) was handed over to the PLAAF. On 8 May 2018, it was announced by PLAAF spokesperson Shen Jinke that Y-20 had "recently conducted its first joint airdrop training with the country's airborne troops.[18] There are rumours about an aerial refuelling variant of the Y-20. China has also reportedly been working on a medium transport aircraft called 'Y 30'. It is likely to be of the C-130 'Hercules' class (20 ton).

Artist's Version of H-20 Stealth Bomber
Picture Source: The National Interest

Civil Aviation Aircraft

China has made major progress in civil aircraft. A C919 airplane took off at the Shanghai Pudong International Airport for a test flight to Xi'an on November 10, 2017.[19] The C919 is a narrow-body twin-jet engine airliner developed by Chinese aerospace manufacturer, Commercial Aircraft Corporation of China (COMAC). The aircraft's maiden flight was on 5 May 2017. Its first commercial deliveries are expected in 2021 to Chinese Eastern Airlines. The aircraft can carry 156 to 168 passengers in a normal operating configuration up to 5,555 km. It is intended to compete primarily with the Boeing 737 MAX and Airbus A320neo. As of 31 August 2018, COMAC has 1,008 commitments including 305 firm orders, mostly from Chinese leasing companies or airlines. The COMAC ARJ21 Xiangfeng (Soaring Phoenix) is a twin-engine regional jet. The 78 to 90 seat jet has a range up to 3,700 km. It made its maiden flight in June 2016 and mass production started in September 2017. Around 30 ARJ21s are in service. In June 2018, an ARJ21-700+ was planned to cut weight and drag for a 115-seat -900 stretch variant to be ready by 2021. COMAC has secured an order for 105 ARJ21-700 regional jets from China's three largest state-owned airlines.[20] The China-Russia joint CR929 wide-body, 250 to 320 seater passenger aircraft has entered the preliminary design phase, according to COMAC.[21] It is meant to challenge the Airbus and Boeing duopoly. A joint venture of COMAC and Russia's United Aircraft Corp., was set up in May 2017 to jointly develop the CR929. More importantly, China's domestic civil aviation industry capacity has been upgraded following the development of the three major models. The next developmental step would be a 400-seat C939. Meanwhile an Airbus A320 plant in Tianjin, China, has been producing aircraft for Chinese airlines since 2009.

Combat Helicopters

The Changhe Aircraft Industries Corporation (CAIC) Z-10 is an attack helicopter designed primarily for anti-tank missions but has secondary air-to-air capability as well. It was co-designed by Kamov design bureau of Russia under a contract with the Chinese government.[22] The Z-10 attack helicopter has recently been fitted with new engine exhaust outlets. Z-19 is the smaller counterpart Harbin Z-10. In September 2016, the PLA announced that all of its ground force aviation units had been equipped with the Z-10. The aircraft is armed with HJ-10 Air-to-Ground Missiles (AGM) considered to be similar to AGM-114 Hellfire and has an anti-helicopter capability in addition to anti-tank capability. The Changhe Z-18 is a medium transport helicopter developed by CAIC to replace the older Z-8. A naval version of the Z-18 was first revealed onboard the Chinese aircraft carrier Liaoning.[23] Earlier in 2014, the army version of Z-18 was revealed in Chinese army publications. The Z-18 has a distinctive boat hull lower fuselage with a tail ramp for offloading small vehicles. It also has a glass cockpit, and extensive use of composites. It has a maximum take-off weight of 13.8 tonnes and can carry 27 troops or five tonnes of cargo with a range of up to 1,000 km. Several versions are planned, including utility/VIP transport, airborne early warning (AEW), and anti-submarine warfare (ASW). It has a chin-mounted surface search radar and is probably fitted with a data-link similar to the Ka-28. The Z-18J AEW variant has a radar antenna fitted in place of the loading ramp that is lowered for operation when airborne. It can operate in high-altitude areas such as Tibet as in flight tests it was able to reach as high as 9,000 m.

Advanced Trainer Aircraft

The Hongdu L-15 Falcon[24] is a supersonic advanced training and light combat aircraft developed by the Nanchang-based Hongdu Aviation Industry Corporation. It will meet the lead-in fighter trainer (LIFT) requirements of PLAAF and People's Liberation Army Naval Air Force (PLANAF), as well as next-generation training and light combat needs for foreign customers. Zambia has already ordered it and Pakistan, Ukraine and Uruguay are in discussion.

Hongdu L-15 Falcon.
Picture Source: AirTeamImages.com

UAVs

There are more than 300 UAV developers/manufacturers in China. China is also called the 'Wal-Mart' of UAVs, as it produces nearly 50 per cent of UAVs used by hobbyists. Over two-thirds of them are private enterprises, and the remaining are government owned enterprises (GOE). GOEs are engaged in the entire development of UAVs of various sizes, from the initial design to the final production in various sizes, ranging from the smallest micro air vehicles to the largest UAVs. The Chengdu Pterodactyl,[25] also known as Wing Loong, is a MALE UAV intended for use as a surveillance and aerial reconnaissance platform. The Pterodactyl I is capable of being fitted with air-to-surface weapons for use in an unmanned combat aerial vehicle (UCAV) role. It can reportedly carry the Chinese BA-7 air-to-ground missile, YZ-212 laser-guided bomb, YZ-102A anti-personnel bomb and 50-kilogram LS-6 miniature guided bomb. There are six variants of Wing Loong. Among the latest are the upgraded Wing Loong II, which can reportedly carry up to 12 air-to-surface missiles. It officially entered service with the PLAAF in November 2018. The Wing Loong ID has improved aerodynamics and engine enabling greater take-off weight, service ceiling, and endurance. It can carry both internal and external stores, as well as communications equipment. The variant was launched in 2018 with Egypt being the first buyer of 32 systems. The variant achieved its first flight on 23 December 2018. Many countries have acquired Chinese UAVs. In October 2018, it was announced that Pakistan Aeronautical Complex (PAC) and Chengdu Aircraft Corporation would jointly produce 48 Wing Loong II UAVs for use in the PAF.[26]

Aerospace Craft

A Chinese hybrid space plane could reset the 21st century space race. It would take off from a runway, fly at hypersonic speeds, and then rocket into orbit.[27] While SpaceX is making news with its recoverable rockets, the Chinese aerodynamically optimized aircraft-cum-space plane would place China ahead of the others in the air and space race by 2030. The China Aerospace Science and Technology Corporation (CASTC) is beginning advanced research on a high tech, more efficient successor to the retired Space Shuttle, with hybrid combined cycle engines that can take off from an airport's landing strip and fly straight into orbit. The hybrid space plane's combined cycle engines would use turbofan or turbojet engines to take-off horizontally from a landing strip. Once airborne, the engine then shifts to ramjet propulsion and, as speed increases, adjusts into a scramjet engine with supersonic airflow. At the scramjet stage, the hybrid space plane would enter hypersonic flight in 'near space', the part of the atmosphere between 20 km and 100 km above sea level. Finally, the hybrid space plane would use its rocket motors to push out of near space and into orbit. The CASTC space plane's easy reusability would exponentially bring down space launch costs. The combined cycle engine shares the same inlet and exhaust nozzle for both the turbojet/turbofan and ramjet. The hybrid spacecraft combined cycle engine and rocket motor prototype will be ready for testing by 2021. The space plane would improve ease of access to space for untrained persons, as it would have more gradual acceleration than a space launch rocket and could be used for space tourism. China is thus the second nation in the world after the USA to master scramjet technology.

Hypersonic Technology

China has the world's largest hypersonic wind tunnel, the Mach 9 JF-12, called the 'Hyper Dragon' (operational since 2014), that could be used to easily test hypersonic scramjets without costly and potentially dangerous flight testing at altitude.[28] The newer FD-21, a 556-foot-long wind tunnel, was finished in 2016 by the China Academy of Aerospace Aerodynamics. It can handle speeds of Mach 10-15, and is large enough to test full-sized components of hypersonic propulsion like gliders and scramjets. NASA's hypersonic wind tunnel reaches to only Mach 7. Clearly, China is not content to restrict its flight research to the lower end of the hypersonic speed range.

The Chinese are also putting in the public domain, designs of a hypersonic military aircraft with a streamlined fuselage, a large ventral air intake, and relatively small wings.[29] Flying in near space (20 km to 100 km altitude), it will be able to dodge existing air defences. This hypersonic bomber could carry a payload of eight tons. Timelines and the success matrix of the project are still evolving. Whether the project meets with success within the targeted timeframe or it slips by several years, only time will tell. The Chinese military could draw direct benefits from the combined cycle engine technology. Without the rocket motor component, the combined cycle engine would be a good fit to power hypersonic UAVs and staffed aircraft. Flying in near space at speeds above Mach 5, such aircraft could have a global reach, while their speed and high altitude would make them effectively immune to all existing air defence systems.

The Shadow Dragon hypersonic bomber concept is envisioned as PLAAF's future arsenal.[30]

The high profile broadcast of Chinese breakthroughs in space technologies suggests that, in addition to boosting Chinese prestige, the Chinese leadership is looking to raise public awareness and support to justify costly investments in next generation space technology such as hybrid space planes and super heavy "Moon" rockets.[31] Getting the first mover advantage in these aerospace milestones would definitely give China a superpower status in both Earth and space. How China and the USA compare in the hypersonic arms race will decide who will control the planet. If Turbine Based Combined Cycle (TBCC) propulsion system succeeds, that would mean China could fly Mach 6 airplanes anywhere in the world in under three hours. American and Chinese defence giants are moving quickly to reach the edge of space. The world is watching this arms race that has the potential to revolutionize access to space and transform transcontinental travel for civilians. It could also transform national security, making existing air and missile defences obsolete.

In the race are two Chinese companies, Chinese Aerospace Science and Technology Corporation (CASC) and Chinese Aerospace Science and Industry Corporation (CASIC), and America's Lockheed Martin, which claims that its hypersonic technology is nearly ready for flight testing. Lockheed Martin's turbine rocket combined cycle (TRCC), an engine that switches between turbofan, ramjet, and scramjets for subsonic, supersonic, and hypersonic flight, will be tested on a fighter-sized flight test bed by 2020. Thereafter, the TRCC will power the SR-72, a Mach 6, unmanned twin-engine reconnaissance aircraft that is targeted to start flight by 2030.[32] This aircraft is meant to enter highly contested and defended airspace at altitudes beyond the stratosphere (30 to 100 km), using its speed to outrun enemy defences like the SR-71 Blackbird once did. Executive Vice President and General Manager at Lockheed Martin Rob Weiss suggested that hypersonic planes could fire hypersonic missiles. This suggests that the SR-72 could join the B-21 stealth bomber in the USAF's future global strike arsenal. Another key U.S. project is the "Phantom Express" XS-1 space plane, which is being developed by Boeing and funded by DARPA. It has recently been closed as Boeing walked out of the project. It is highly likely that because of the nature of material sciences and the laws of physics, hypersonic aircraft like the American SR-72 and its Chinese counterparts would look similar to each other.

The Tengyun space plane, which CASIC intends to deliver by 2030, will use a TRCC-engine hypersonic carrier aircraft to carry the second rocket-powered stage into near space,[33] much like the abandoned XS-1. Both stages will be reusable; the hypersonic carrier airframe could also be used in hypersonic strike and reconnaissance roles. Teng Yun would have a combined weight of 100-150 tons. The first stage is a Mach 6 hypersonic 'carrier' aircraft, with TRCC engines. Unlike the XS-1, the Teng Yun's second stage is a reusable, 10-15-tonne rocket-powered space plane capable of carrying either 2 tons of cargo or 5 passengers. CASIC Vice President Liu Shiquan said that the company has finished technology demonstration of key Teng Yun flight technology like the engine. The two-stage space plane could be flying by 2030, as part of a $16 billion research effort. In theory, it could cross the Pacific Ocean in one hour. Like the U.S. systems, the CASIC hypersonic carrier could have other applications besides space launch. It's big enough to carry significant payloads (10-15 tons), which could include sensors and weapons. With modifications, like additional fuel tanks for increased range, its hypersonic speed and near-space flight altitude could fit with Chinese military needs for fast, global reconnaissance and strike. As CASIC manufactures most of China's missiles, its experience with rocket motors, ramjet engines (and future scramjet engines) means that it is likely that it has the resources to build the complex TRCC engines for the hypersonic carrier, and by extension a hypersonic recon and strike plane.

The CASC's hypersonic space plane, which was announced in August 2016, is also targeting demonstration of critical technologies by 2020, and a full-scale hypersonic space plane is expected to fly by 2030.[34] The turbo-aided rocket-augmented ram/scramjet engine (TRRE), which uses rocket augmentation in order to aid in the transition into the supersonic and hypersonic flight regimes, could be the world's first combined cycle engine to fly in 2025, paving the way for hypersonic near-space planes and single-stage space launchers. Further improvements will lead to a scramjet engine powerful enough for the hypersonic plane to fly directly to orbit in one piece, as compared to the two-stage model. The Variable Specific Impulse Magneto-plasma Rocket (VASIMR) is an electro-thermal thruster under development for possible use in spacecraft propulsion.[35] It will use radio waves to ionize and heat propellant. Then a magnetic field accelerates the resulting plasma to generate thrust. Presumably, CASC's interplanetary nuclear spaceship will operate on similar principles. CASC also has equally ambitious plans for deep-space exploration. Key among these plans is for a nuclear-powered spacecraft, which is slated to enter service in 2040, most likely to support a manned Chinese Martian mission.[36] CASC also has plans for China to engage in deep-space economic activity, like building orbit solar power plants, and mining asteroids and the moon.

PLAAF's Hypersonic military aircraft would have a streamlined fuselage, a large ventral air intake, and relatively small wings. Flying in near space at speeds of up to Mach 10, it will dodge all existing air defences. It has the potential to revolutionize both military and civilian aerospace. Chinese engineers are working on a wide range of other hypersonic technologies such as plasma jets to steer hypersonic thrust, advanced heat resistant composites, and new fuels. China is set for a hypersonic flight boom.

Advances in Space Technology

CASC's Shanghai Academy of Spaceflight Technology will start testing vertical landing technology for LM-4 and LM-6 rockets in 2019 and 2020.[37] To launch the thousands of tons of payload required to execute these plans, CASC is planning to make its rockets all completely reusable by 2035, from the small Long March 6 all the way to the giant Long March 5. The Long March 4B tested grid fins in 2019[38] (grid fins on reusable rockets like the SpaceX Falcon are used to control flight during the high velocity portion of the descent phase) and vertically land a Long March 7 rocket in 2020.[39]

Since 2003, when the country launched its first astronaut, the multibillion-dollar space program run by the Chinese military has been right on schedule with achieving the landmarks it set for itself. The five-year plan released by Beijing's State Council Information Office in November 2017 stresses that China is committed to peaceful uses of space and seeks to avoid a space arms race. While the USA and Russia have historically dominated space research and exploration, China's military-backed space program landed a roving vehicle 'Chang 3' on the moon in 2013, making China the third nation (after the USA and the former USSR) to reach the moon. Chang 3 discovered a new type of basaltic rock during its exploration of a volcanic crater in the Mare Imbrium.[40] In November 2016, two Chinese astronauts concluded a 30-day mission aboard the country's own space station Tiangong 2.[41] China's ambitions for its space program include a soft landing on the moon, a probe to Mars and private foreign investment. On 03 January 2019, the Chinese robotic spacecraft

'Chang 4' made a soft landing in the South Pole of the Aitken Basin area of the moon, otherwise known as the "far side" or "dark side" of Earth's only natural satellite.[42] Till now, China was emulating space milestones already reached by the Americans and Russians. It is the first spacecraft in history to attempt or achieve a landing on this unexplored area, which is never visible from Earth. The basin itself is the largest known impact crater on the moon, and one of the largest in the entire solar system. The distance from its depths to the tops of the highest surrounding peaks measures some 15 km, almost twice the height of Mount Everest.

Space will remain an important part of President Xi's plan to make China an 'earthly paradise' by 2050. Astronaut Jing Haipeng described China's planned space station 'a glorious mission'.[43] China will conduct an average of 30 launches per year by 2020, and overtake the USA in some 'key aerospace projects' by 2045.[44] The Chinese aerospace industry has labelled a 'strategic emerging industry', along with oceanography, information networks, life sciences and nuclear technology. The Chinese Communist party White Paper on space presented in the 19th Party Congress elaborates that China's space programme will first work on "high-thrust liquid oxygen and kerosene engines, and oxygen and hydrogen engines", for heavy-lift launch vehicles. China's next-generation heavy-lift rocket, the 'Long March 9', is currently under development. The rocket will be able to take 140,000 kg to Low Earth Orbit (LEO) and 50,000 kg to Lunar Transfer Orbit (LTO).[45] There is a mention of developing non-toxic and pollution-free medium-lift launch vehicles, with the general aim of reducing pollution. There is a plan to develop small, reusable launch vehicles.

BeiDou Navigation Satellite System

The second generation space navigation system, BeiDou Navigation Satellite System (BDS) became operational in China in December 2011 with a partial constellation of 10 satellites in orbit. Since December 2012, it has been offering services to customers in the Asia Pacific region.[46] In 2015, China started the build-up of the third generation BeiDou system (BeiDou-3) in the global coverage constellation. The first BDS-3 satellite was launched on 30 March 2015. On 27 December 2018, the BeiDou Navigation Satellite System started providing global services. Project Director Ran Chengqi reported that the core of the BDS was completed in December 2019. BeiDou-3 now consists of 35 satellites and is expected to provide global services in 2020.[47] BeiDou will provide an alternative global navigation satellite system to the USA-owned Global Positioning System (GPS), and the Russian GLOSNASS or the European Galileo systems and is expected to be more accurate than them, it is claimed, with millimetre-level accuracy. According to the government mouthpiece, *China Daily*, in 2015, fifteen years after the satellite system was launched, it was generating a turnover of $ 31.5 billion per annum for major companies such as China Aerospace Science and Industry Corp, AutoNavi Holdings Ltd, and China North Industries Group Corp.[48]

Tiangong Space Station

Tiangong-2, China's second space laboratory module, was launched on 15 September 2016. It carries a total of 14 experiment packages, including Space-Earth quantum key distribution and laser communications experiment to facilitate space-to-ground quantum communication.[49] The satellite, nicknamed Micius after the ancient Chinese philosopher and scientist, has successfully conducted quantum entanglement and established a 2000-km quantum communication link between Beijing and Austria.[50] Experiments over long distances allow the development of quantum encryption and quantum teleportation technology to facilitate communications safe from eavesdropping. By producing pairs of entangled photons, it allows ground stations separated by many thousands of kilometres to establish secure quantum channels. More Micius satellites will follow, allowing a European–Asian encrypted network by 2020, and a global network by 2030.[51] In 2017, China's first cargo spacecraft, Tianzhou-1, completed automated docking and refuelling with its Tiangong-2 space lab, showing the country's progress in running its own space station. The frontier fields and experiments to be conducted in space include research into dark matter using the 'hard X-ray modulation telescope - HXMT' launched in June 2017. The next great leap forward for China's space program will probably occur in 2020, if it succeeds in accomplishing three major tasks – a mission to Mars, completion of a space station, and in-orbit testing of a space-based solar power system that can beam energy back to Earth.[52]

Tiangong-2 Space Laboratory. Image: China Manned Space Engineering Office, Via spaceflight101.com

Long March 9 Launch Rocket

Capable of carrying 140 tons to low earth orbit (or 50 tons on a lunar bound course), the Long March 9 will be the world's most powerful space launch rocket when it flies around 2030.[53] It'll be the workhorse not just for lunar missions, but for trips to Mars and beyond. The Long March 9 will likely be used for manned missions to the moon as well as installing orbital solar power plants. Its massive payload capacity would also support Chinese plans to explore Jupiter, Saturn and its moons, and other celestial objects beyond the asteroid belt. The Long March 9's huge payload could come in handy for creating a base on the moon as well. It is likely to be a robotic facility; a manned base would cost a whole lot more. Jiao Weixin of Peking University says that a robot base dug into the lunar surface, unlike lunar rovers and landers, could conduct more sophisticated scientific tests of lunar soil, and send back lunar rock samples cheaply.[54] Researching and operating all this is going to be very expensive, but it is clear China views the new space race as a way to pick up a lot of prestige. These new plans will be added to China's already wide range of space offerings, including navigation and spy satellites, a vibrant hypersonic technology industry, and a growing robotics and AI industries.

Reusable Space Planes

China has been making some ambitious space plans backed by lots of money. These include reusable space planes, nuclear-powered spaceships, and robotic moon bases. In response, President Trump signed a Presidential

Memorandum in 2018 directing NASA to send astronauts back to the moon for long-term exploration and to prepare for the long-awaited manned Mars mission.[55] Billionaires Elon Musk, Jeff Bezos, and Richard Branson dominate the emerging industry of commercial spaceflight. They are competing to put satellites or tourists, or both in space in their bids to become cheaper alternatives to NASA, the European Space Agency, and other government-run space programs. However, like pioneers in the smart-phone and artificial intelligence industries before them, they now face competition from a deep-pocketed China that threatens to disrupt their launch-pads. President Xi Jinping has loosened the government's monopoly on space launches, and that's fuelling the formation of small domestic companies with ambitions to challenge Musk's SpaceX, Bezos' Blue Origin, and Branson's Virgin Galactic.[56] They can also rely on the expertise of rocket scientists from China's vigorous manned space program. "We are really a start-up growing on the shoulders of the state aerospace giant," says Zhang Changwu, chief executive officer of Beijing-based Landspace Technology Corp.[57] "There's no better time for a commercial rocket firm to grow in China than now." The number of satellites in space increased 50 per cent from 2013 to 2017, to 1,738, according to the Satellite Industry Association. Chinese launchers could help manufacturers seeking an inexpensive way to get even more of them into orbit. One lucrative payload for the Chinese could be miniature satellites, which can weigh less than 200 pounds. Launching these satellites will generate $15 billion in revenue by 2027, up from less than $100 million last year, according to the Paris consulting firm Euroconsult.[58] That's also a niche where big players such as Space Exploration Technologies Corp. (SpaceX), whose Falcon Heavy rocket can lift into orbit payloads of almost 64 metric tons (141,000 pounds), have been less active. Several mainland companies have succeeded with suborbital launches and are vying to be the first in China to place satellites into orbit around the Earth. Founded in 2015, Landspace had raised 500 million Yuan ($ 72 million) from local investors by April 2018 and employs 170 rocketeers and other engineers, almost all veterans of the national space program, Zhang says.[59] The company plans to launch a rocket this month carrying a satellite for state-run broadcaster CCTV. Two other Chinese companies have orbital launches scheduled by 2020. More than 60 Chinese companies have entered the commercial space industry in the past three years, the state-run Xinhua News Agency reported in May.

China is the second-largest market for Tesla Inc.'s electric vehicles and has inspired many start-ups. "SpaceX has had a huge impact," says Lan Tianyi, founder of Ultimate Blue Nebula, a Beijing-based space consultant. "Nobody in China thought that a private company could develop a rocket. Now they see that a U.S. company can do this, and it's very famous in China." China wants to be one of the world's top three aerospace powers by 2030, says Wu Yanhua, vice administrator of the China National Space Administration. Toward that goal, Xi envisions private-sector launches of satellites for customers in China and in countries participating in the 'Belt and Road' initiative and its program to build ties throughout Africa, Asia, and Europe by funding infrastructure projects. Landspace opened a factory costing almost 500 million Yuan about 90 miles west of Shanghai in 2018, with capacity to produce 15 rockets using liquid propellant annually. The facility eventually could make 200 engines a year, Zhang says.[60] Another start-up, Beijing Interstellar Glory Space Technology Co., in September 2018 sent three test satellites into space aboard a solid-fuel rocket. The company, founded in 2016 and also known as i-Space, has raised 600 million Yuan from investors such as Shunwei Capital, whose chairperson, Lei Jun, is co-founder of Smartphone maker Xiaomi Corp. Blaine Curcio, founder of Orbital Gateway Consulting in Hong Kong, says most Chinese rocketeers have a long way to go because their rockets aren't reusable and can't handle heavier payloads.[61] Instead, the start-ups will—for now, at least—count on demand from companies developing smaller, lighter satellites for meteorological, telecommunications, and navigational use. With an $ 8 billion national space budget, China wants to build a commercial launch industry to join the ranks of Western leaders such as SpaceX, Blue Origin, and Virgin Galactic.

China Catch Up Strategy: Key Technologies

China is looking to catch up with rivals like the USA and Germany in high-end technology, making a major push with a "Made in China 2025" strategy that identifies 10 key sectors, including robotics, aerospace and clean-energy cars. President Xi Jinping's drive is to neutralize the effects of the trade standoff between China and the USA. Chinese trade partners in Europe, especially Germany, have also raised concerns that a more protectionist China is aggressively moving up the value chain faster than expected. Among them are some key aviation related targets. Chinese plane makers are required to capture 10 per cent of the domestic market. The home-grown CJ-1000A jet engine should also be ready for commercial use.

In the space race, China wants 80 per cent of civil space industry equipment to be domestically sourced by 2025.[62]

American Assessment of Chinese Aerospace

The U.S. Defence Intelligence Agency (DIA) report released on January 15, 2019 suggested that "China's military rise is well-planned, and Chinese leaders are following a strategy they believe will lead to greater power and influence, both regionally and globally...".[63] The report is a U.S. perspective on the current and evolving state of China's military and its ability to wage war. Space operations, Chinese military thinkers thought, were a critical aspect of future warfare. 'Space operations probably will form an integral component of other PLA campaigns and serve a key role in enabling actions to counter third-party intervention during military conflicts,' the report contends. "The China Military Power report emphasizes the PLA's growing reliance on a range of satellite applications for its operations, to include Earth observation, communications, and positioning, navigation, and timing (PNT) satellites, much like the military of its U.S. rival. China Military Power, also highlights, however, Beijing's continuing efforts in developing counter-space capabilities and anti-satellite weapons." A 12 February 2019 article in *Japan Times* also reported that "China is developing sophisticated space capabilities such as 'Satellite inspection and repair' and debris cleanup—at least some of which could also function as weapons against U.S. satellites," according to the Defence Intelligence Agency.

American Response

The "U.S. military will withdraw hundreds of troops focused on counter terrorism operations in Africa over the next several years to support the Pentagon's increased focus on countering threats from China and Russia," officials said.[64] Air Force Secretary Heather Wilson had called for increasing the Air Force from its current size of 312 operational squadrons to 386 by 2030, as it prepares for a possible conflict against a major nation such as China or Russia.[65] This 24 per cent increase in squadrons is the centre-piece of the service's proposal, which has been in the works for six months. This proposal seeks to lay out what it would take for the Air Force to fight a peer adversary and win, as well as defend the homeland, provide a credible nuclear deterrent, counter a medium-sized rogue nation that might try to take advantage of the Air Force's focus on the major adversary, and fight violent extremists such as the Taliban and the Islamic State. "We must see the world as it is," Wilson said. "That was why the National Defence Strategy explicitly recognizes that we have returned to an era of great power competition." Effectively, the USAF intends to add 70 new squadrons to counter rising China and muscle-flexing Russia. It will also mean 40,000 additional personnel, says Lt. Gen. Brian Kelly, commander of the Air Force Personnel Centre. A coalition between Russia and China is highly possible, the USA fears.

Chinese Muscle Flexing

President Xi Jinping told his military commanders to "concentrate preparations for fighting a war" as tensions continue to grow over the future of the South China Sea and Taiwan. "We need to take all complex situations into consideration and make emergency plans accordingly," President Xi, told the officers of the Southern Theatre Command.[66] President Xi's words represent a significant ramping-up of the rhetoric being exchanged between Beijing and Washington. China had been angered by U.S. sanctions on its military for buying weapons from Russia, and by what Beijing sees as renewed Washington support for democratic Taiwan. President Xi "stressed the need to focus on combat research and commanding, to advance work in all areas and accelerate developing strong and efficient joint-operation commanding institutions for theatre commands to comprehensively boost the military's battle-winning ability."[67]

China's Aerospace Power: Realities India

The strength of China's PLA has dramatically increased in recent years. PLAN now has more warships and submarines than the USA. The USN remains technically more advanced with many powerful units including their nuclear-powered aircraft carriers. That quality gap, however, is rapidly closing. China has brought into service with surprising speed new stealth fighters and long-range missile-carrying bombers. Its warships are now equipped with advanced radars and control systems. Two relatively small aircraft carriers are being used to test and practise naval aviation warfare, and military analysts believe a further four carriers currently being built will be much larger and possibly nuclear powered.

Pushan Das of the Observer Research Foundation wrote about the widening gap between Indian and Chinese air power on 30 April 2018.[68]

"As the Chinese PLA continues to undertake significant organizational and modernization reforms to build a military that is 'built to fight', in line with President Xi Jinping's recent rhetoric at the 19th Party Congress, asymmetric inequalities between the IAF and the PLAAF are set to widen". If the IAF is to attain the capability to counter and fight any future informatised war or conflagration vis-à-vis the PLAAF, New Delhi must re-evaluate its existing and future air combat capability, Das adds. China's progress in upgrading its air combat capability has been considerable in terms of the expansion of its aircraft inventory and extended range air-to-air missiles. The IAF's combat enablers like FRA and AEW&C could be targeted by long-range missiles like the PL-15. The PLAAF views stealth technology "as a core capability in its transformation from a predominantly territorial air force to one capable of conducting both defensive and offensive operations", according to the U.S. Department of Defence. China's fifth-generation fighter jets, represent a significant threat to the IAF. The PLAAF-IAF capability gap will continue to widen due to underfunded modernization and very slow procurement process. Pushan further quotes Air Vice Marshal (Retd) Arjun Subramaniam who has argued that the ability of the latest variant of the Chinese-built H-6 bomber to carry six DH-10 cruise missiles—each with a range of around 1,500 km and a combat radius of 1,800 km—means that "Chinese stealth fighters and bombers do not need to become airborne from airfields in Tibet but can launch their cruise missiles on critical Indian military targets from well outside any kind of air defence umbrella the IAF can put in place over the next decade." The IAF currently is at an all-time low of 29 fighter squadrons. The 36 Rafale and very slowly inducting Light Combat Aircraft (LCA), or even acquisition of additional SU-30 MKIs will not be able to stem the further decline. India's indigenous fifth generation fighter, the Advanced Medium Combat Aircraft (AMCA), is still over a decade away. Increased defence budgeting and faster indigenization are the only alternatives for India to match the aggressively growing aerospace might of China.

IAF Air Chief Marshal B.S. Dhanoa, in a first-ever frank assessment of the northern neighbour's activities, had described China's PLAAF as one of the fastest growing air forces in the world, ranked second in terms of combat airpower.[69] Air Chief Dhanoa also said that during the process of modernization the PLAAF has developed credible year-round capability in the world's toughest terrain—the Tibet region. "Over the past few years, we have seen a significant increase in aircraft and aircrew in the TAR from other military regions. The deployment of Sukhoi-27 and J-10 fleets (in Tibet) for continuous operations during the winter months affords them a credible year-round capability. Earlier, many years ago, they only used to occupy the airfields during the summer," Dhanoa said. During a talk on the "Role of IAF in the Changing Security Environment" at the Vivekananda International Foundation (think tank) in New Delhi Dhanoa said, "It (China) has a credible mix of multi-role fighter and strike aircraft. They have got an adequate reserve to replenish after attrition. They have multi-layered air defence systems and rocket forces that allow them to fight a ground campaign without a decisive air victory."[70] He further added that China had developed a modern air force that relies on quality rather than quantity and ensures that more than 50 per cent of its fleet will comprise advanced multi-role combat aircraft. The IAF Chief's statement comes after the force, that is currently facing a shortage of over 200 fighter jets, had just conducted an exercise named 'Gagan Shakti'. The exercise was aimed at testing the IAF's 'Plan B', which it intends to press into action in the event of a "two-front war" situation.

Rajesh Rajagopalan, a professor of international politics at Jawaharlal Nehru University in New Delhi, has looked at the strategic choices with India.[71] China's alignment with Pakistan and deepening relations with other South Asian countries represent a significant challenge to India's position in the region. Beijing's ability to provide financial assistance and balance against New Delhi may tempt India's smaller neighbours to play one power against the other, undermining India in its own backyard. A strategy of non-alignment, hedging, or alignment with China would not serve India's interests because China's power, geographical proximity, and policies already represent a clear danger to India's security and global interests. A closer alignment with the USA, along the same policy path that India is already pursuing, represents the best way to meet the challenge of China's rise, because the USA is the only power that is stronger than China. Further, New Delhi and Washington share a common interest in balancing Beijing. Among India's other strategic options, efforts to build indigenous military power and forge regional partnerships are necessary and complementary means of countering China, but are by themselves insufficient, because China is already wealthier and stronger than India or any combination of other Asian powers, he writes.

However, the IAF and India's space program have been heavily dependent on Russian technology and systems. Nearly 60 per cent of the

IAF is composed of Russian aircraft. The SU-30 MKI fleet is nearly 40 per cent of the total IAF fighter assets. The worsening US-Russia relations and application of the U.S. law, Countering America's Adversaries Through Sanctions Act (CAATSA) creates complications for the two countries. The USA is reluctant to part with high end aerospace technologies when India continues to buy modern Russian weapon systems such as the S-400 air defence system. If India wants to be a global player, it has to invest more resources in indigenous aerospace technologies. In the meanwhile, India may have no choice but to continue to do the balancing act between the USA and Russia, and continue to engage China.

REFERENCES

1. Chairman Marco Rubio, US Senate Committee on Small Business and Entrepreneurship, Made in China 2025 and the future of American Industry, 2019, https://www.rubio.senate.gov/public/_cache/files/d1c6db46-1a68-481a-b96e-356c8100f1b7/3EDECA923DB439A8E884C6229A4C6003.02.12.19-final-sbc-project-mic2025-report.pdf
2. Ibid, p. 25.
3. Annual report to US Congress, Military and Security Developments Involving the People's Republic of China 2019, Office of the Secretary of Defence, Executive Summary, May 02, 2019, https://media.defense.gov/2019/May/02/2002127082/-1/-1/1/2019_CHINA_MILITARY_POWER_REPORT.pdf
4. Ibid.
5. Ibid.
6. Ibid.
7. Ibid, p. 40.
8. Ibid, p. 41.
9. Bloomberg, The Chinese missiles changing the game in the air and the global arms trade, *South China Morning Post*, May 8, 2018, https://www.scmp.com/news/china/diplomacy-defence/article/2145115/chinese-missiles-changing-game-air-and-global-arms
10. Ibid.
11. Alex Hollings, Counterfeit Air Power: Meet China's Copycat Air Force, *Popular Mechanics*, September 19, 2018, https://www.popularmechanics.com/military/aviation/g23303922/china-copycat-air-force/
12. Ibid.
13. Ibid.
14. Jon Lake, J-20 Mighty Dragon Enters Frontline Service, *Asian Military Review*, January 28, 2018, https://asianmilitaryreview.com/2019/01/j-20-mighty-dragon-enters-frontline-service/
15. Shenyang FC-31, From Wikipedia, the free encyclopaedia, https://en.wikipedia.org/wiki/Shenyang_FC-31
16. Sebastian Roblin, Danger: Should We Fear China's New H-20 Stealth Bomber? *The National Interest*, January 3, 2020, https://nationalinterest.org/blog/buzz/danger-should-we-fear-chinas-new-h-20-stealth-bomber-110681

17. Liu Yang and Li Qiang, Aviation Video Suggests China has Developed new Long-range Bomber, *Global Times*, May 9, 2018 http://www.globaltimes.cn/content/1101430.shtml
18. Deng Xiaoci, China's homegrown transport plane Y-20 conducts 1st air drop drill, *Global Times*, May 8, 2018 http://www.globaltimes.cn/content/1101274.shtml
19. Zhu Wenqian, C919 completes long-distance flight, *China Daily*, November 11, 2011, https://www.chinadaily.com.cn/business/2017-11/11/content_34392789.htm
20. Aaron Hilsz Lothian, COMAC secures order for 105 arj21-700 jets, Samchui Aviation and Travel, August 31, 2019, https://samchui.com/2019/08/31/comac-secures-order-for-105-arj21-700-jets/#.XmjcsXIzbIU
21. Zhao Lei, Aircraft firm reports progress in China's joint jetliner project, *China Daily*, Updated September 05, 2019, https://www.chinadaily.com.cn/a/201909/05/WS5d706338a310cf3e35569d2e.html
22. Staff Reporters, CAIC Z-10 (Fierce Thunderbolt), Military Factory, https://www.militaryfactory.com/aircraft/detail.asp?aircraft_id=600
23. Z-18 Medium transport helicopter, *Military Today*, http://www.military-today.com/helicopters/z18.htm
24. Hongdu L-15 Supersonic Trainer/Attack Aircraft, Air Force Technology, https://www.airforce-technology.com/projects/hongdusupersonictrai/
25. CAIG Wing Loong, Wikipedia, https://en.wikipedia.org/wiki/CAIG_Wing_Loong
26. China to sell 48 high-end military drones to Pakistan, *Economic Times*, October 10, 2018, https://economictimes.indiatimes.com/news/defence/china-to-sell-48-high-end-military-drones-to-pakistan/articleshow/66129500.cms?from=mdr
27. Jeffrey Lin and P.W. Singer, China's Hybrid Spaceplane Could Reset The 21st Century Space Race, *Popular Science*, August 9, 2016, https://www.popsci.com/chinas-hybrid-spaceplane-could-reset-21st-century-space-race/
28. Jeffrey Lin and P.W. Singer, A look at China's most exciting hypersonic aerospace programs, *Popular Science*, April 18, 2017, https://www.popsci.com/chinas-hypersonic-technology/
29. Ibid.
30. Jeffrey Lin and P.W. Singer, China's Hybrid Spaceplane Could Reset The 21st Century Space Race, *Popular Science*, August 9, 2016, https://www.popsci.com/chinas-hybrid-spaceplane-could-reset-21st-century-space-race/
31. Ibid.
32. Jeffrey Lin and P.W. Singer, American and Chinese aircraft could be flying 4,000 miles per hour by 2030, *Popular Science*, June 16, 2017, https://www.popsci.com/hypersonic-arms-race-china-united-states/
33. Ibid.
34. Ibid.
35. Variable Specific Impulse Magnetoplasma Rocket (VASIMR), Wikipedia, https://en.wikipedia.org/wiki/Variable_Specific_Impulse_Magnetoplasma_Rocket
36. Jeffrey Lin and P.W. Singer, China could become a major space power by 2050, *Popular Science*, December 18, 2017, https://www.popsci.com/china-space-power-plans/
37. Jeffrey Lin and P.W. Singer, Here's China's plan to compete with SpaceX and Blue Origin, *Popular Science*, May 14, 2018, https://www.popsci.com/chinas-2020-plan-for-reusable-space-launch/
38. Rui C. Barbosa, Long March 4B launches Gaofen-7 – tests grid fin system, Nasa Space Flight, November 3, 2019, https://www.nasaspaceflight.com/2019/11/63864/

39. Andrew Jones, China prepares to launch new rockets as part of push to boost space program, Space Com, February 2020, https://www.space.com/china-space-program-new-long-march-rockets.html
40. Danny Lewis, China's Lunar Rover Discovered a New Kind of Moon Rock, Smithsonianmag.com, December 24, 2015, https://www.smithsonianmag.com/smart-news/chinas-lunar-rover-discovered-new-kind-moon-rock-180957638/
41. Stephen Clark, Two Chinese astronauts back on Earth, Space Flight Now, November 18, 2016, https://spaceflightnow.com/2016/11/18/shenzhou-11-landing/
42. Jeremy Chan, Chang'e-4 Makes Historic Landing On Far Side Of Moon, *Asian Scientist Magazine*, January 4, 2019, https://www.asianscientist.com/2019/01/topnews/chang-e-4-far-side-moon-historic-landing/
43. Deyana Goh, China's upcoming missions according to its 5-Year Plan & Space White Paper, Space Tech Asia, October 29, 2017, https://www.spacetechasia.com/chinas-upcoming-missions-according-to-its-5-year-plan-space-white-paper/
44. Ibid.
45. Ibid.
46. Nineteenth meeting of the communications/navigation and surveillance sug-group (cns sg/19) of Apanpirg Bangkok, Thailand, International Civil Aviation Organization, 20 – 24 July 2015, https://www.icao.int/APAC/Meetings/2015%20CNSSG19/IP18-China%20AI5%20-%20Progress%20of%20the%20BDS%20n%20SARPS.pdf
47. Jonathan Shieber, China nears completion of its GPS competitor, TechCrunch.com, December 28, 2019, https://techcrunch.com/2019/12/28/china-nears-completion-of-its-gps-competitor-increasing-the-potential-for-internet-balkanization/
48. Gao Yuan, Sky's the limit for Beidou's clients, *China Daily*, November 16, 2015, https://www.asiaone.com/business/skys-limit-beidous-clients
49. Rui C. Barbosa, China launches Tiangong-2 orbital module, NASA SpaceFlight.com September 14, 2016, https://www.nasaspaceflight.com/2016/09/china-launch-tiangong-2-orbital-module/
50. BBC News, China launches quantum-enabled satellite Micius, August 16, 2016, https://www.bbc.com/news/world-asia-china-37091833
51. Tomasz Nowakowski, China's 'Micius' Satellite, Space Flight Insider, January 22, 2018, https://www.spaceflightinsider.com/organizations/china-national-space-administration/chinas-micius-satellite-demonstrates-intercontinental-quantum-key-distribution-first-time/
52. Deyana Goh, China's upcoming missions according to its 5-Year Plan & Space White Paper, *Space Tech Asia*, October 29, 2017, https://www.spacetechasia.com/chinas-upcoming-missions-according-to-its-5-year-plan-space-white-paper/
53. Zhao Lei, Super-powerful Long March 9 said to begin missions around 2030, *China Daily* Updated, March 11, 2019 http://www.chinadaily.com.cn/a/201903/11/WS5c859b62a3106c65c34edcc0.html
54. Jeffrey Lin and P.W. Singer, China could become a major space power by 2050, *Popular Science*, December 18, 2017, https://www.popsci.com/china-space-power-plans/
55. NASA, New Space Policy Directive Calls for Human Expansion Across Solar System, Dec. 12, 2017, https://www.nasa.gov/press-release/new-space-policy-directive-calls-for-human-expansion-across-solar-system
56. Bloomberg, INNOVATION: China wants to build the first solar power station in space. February 19, 2019, https://energy.economictimes.indiatimes.com/news/renewable/innovation-china-wants-to-build-the-first-solar-power-station-in-space/68059211

57. Bloomberg Quint, China's Space Program Is Coming for Elon Musk and Jeff Bezos, October 23, 2018, https://www.bloombergquint.com/businessweek/china-s-space-program-is-coming-for-elon-musk-and-jeff-bezos
58. *Business Standard*, Bloomberg, China's space programme is coming for Elon Musk and Jeff Bezos, Last Updated at October 19, 2018, https://www.business-standard.com/article/international/china-s-space-programme-is-coming-for-elon-musk-and-jeff-bezos-118101900015_1.html
59. Bloomberg Quint, China's Space Program Is Coming for Elon Musk and Jeff Bezos, October 23 2018, https://www.bloombergquint.com/businessweek/china-s-space-program-is-coming-for-elon-musk-and-jeff-bezos
60. Ibid.
61. Ibid.
62. Di Lettura, Factbox - Made in China 2025: Beijing's big ambitions from robots to chips, Reuters, April 20, 2018, https://it.reuters.com/article/idUKKBN1HR1DO
63. Keith K.C. Hui, US-China arms race has begun with space weapons, *China News*, February 13, 2019, https://chinanewsstories.com/2019/02/13/us-china-arms-race-has-begun-with-space-weapons/
64. Reuters, Pentagon to cut hundreds of troops in Africa as U.S. focuses on China, Russia, *First Post*, November 16, 2018, https://www.firstpost.com/world/pentagon-to-cut-hundreds-of-troops-in-africa-as-u-s-focuses-on-china-russia-5559641.html
65. Stephen Losey, Air Force calls for 74 more squadrons to prepare for possibility of war against major power. *Air Force Times*, September 17, 2018, https://www.airforcetimes.com/news/your-air-force/2018/09/17/air-force-calls-for-74-more-squadrons-to-prepare-for-possibility-of-war-against-major-power/
66. Kristin Huang, 'Prepare for war', Xi Jinping tells military region that monitors South China Sea, Taiwan, *South China Morning Post*, October 26, 2018, https://www.scmp.com/news/china/military/article/2170452/prepare-war-xi-jinping-tells-military-region-monitors-south
67. Jamie Seidel, China's warning: Prepare for war, *The Daily Examiner*, October 29, 2018, https://www.dailyexaminer.com.au/news/president-xi-tells-military-to-concentrate-prepara/3561651/
68. Pushan Das, The widening gap between Indian and Chinese air power. ORF, April 30, 2018, https://www.orfonline.org/research/the-widening-gap-between-indian-and-chinese-air-power/
69. Sputnik, Indian Air Force Chief Lauds China's Air Defences in World's Toughest Terrain, April 27, 2018, https://sputniknews.com/asia/201804271063969705-iaf-lauds-chinese-air-defense/
70. Rahall Singh, No import of defence platforms at the cost of indigenisation: Air Force Chief, *Hindustan Times*, April 28, 2018, https://www.hindustantimes.com/india-news/no-import-of-defence-platforms-at-the-cost-of-indigenisation-air-force-chief/story-EuU7X9jIeZbAECMBNUDIRI.html
71. Rajesh Rajagopalan, India's Strategic Choices: China and the Balance of Power in Asia, Carnegie India, September 14, 2017, https://carnegieindia.org/2017/09/14/india-s-strategic-choices-china-and-balance-of-power-in-asia-pub-73108

4

China's Aviation Industry:

Pulling Ahead, Yet Critical Technology Challenges

"When once you have tasted flight, you will forever walk the earth with your eyes turned skyward, for there you have been, and there you will always long to return."

—**Leonardo daVinci**

"The opportunity to secure ourselves against defeat lies in our own hands, but the opportunity of defeating the enemy is provided by the enemy himself."

—**Sun Tzu**

China conducted its biggest ever military parade on 01 Oct 2019 as part of its celebration of 70 years of Communist Party rule. Over 580 pieces of military equipment and 160 aircraft participated.[1] It included the fifth-generation fighter, the J-20; the latest strategic bomber H6-N-cum-FRA; and the Z-20 medium lift helicopter, similar to the U.S. UH-60 Black Hawk. A new advanced radar system that could 'detect jets and missiles', and the latest HQ-9B surface-to-air missiles capable of intercepting multiple air strike weapons in a complex electro-magnetic environment, were showcased among many others.

China's massive military modernisation is being matched by equally phenomenal growth of its military industrial base. The clear focus is to counter the primacy of the U.S. military in the region. For long, China was accused of acquiring technology by reverse engineering Russian systems[2] in clear violation of intellectual property rights, or acquiring top-end technology through cyber-attacks, and espionage operations against Western firms. However, ever since it became an economic powerhouse, it was in a position to invest in research and development. It has clearly focused on platforms that will help them dominate the western Pacific, military aviation, and naval ship building. China's usurping of the SCS, by reclaiming islands and turning them into military bases was a part of the grand strategy to extend the zone of economic and military influence. Air is the preferred means of extending reach with lethality, and even among ships, the crown jewels are the aircraft carriers. As per the London-based International Institute for Strategic Studies, "Since 2014, China has launched more submarines, warships, principal amphibious vessels and auxiliaries than the total number of ships currently serving in the navies of Germany, India, Spain, Taiwan and the United Kingdom."[3]

China's aviation industry has also been leaping ahead. Though the PLAAF is still dominated by the Chengdu J-10 and the Shenyang J-11/15/16 derivatives of the Russian Sukhoi Flanker family, albeit they have significant local avionics and weapons. The Y-9 turboprop and Y-20 jet airlifter are in serial production. Despite virtual disappearance from the news, China's FC-31 second stealth fighter jet is claimed to be proceeding smoothly and on schedule. Chinese armed drones have great potential to dig into Western markets. The Chinese military claims to have tested an unmanned single-engine biplane transport aircraft that successfully delivered cargo at a designated area.[4]

Despite great efforts and investments, the Chinese aircraft industry continues to struggle with critical technologies. China is still struggling to build reliable aircraft engines. China has been researching to develop thrust-vectoring aircraft engines for nearly two decades. At the 12th Zhuhai Air Show in November 2018, a J-10B equipped reportedly with an indigenous thrust vector control nozzle flew and performed aerobatic manouvres.[5] China is scouting to acquire or support cash-strapped engine companies around the world. The Ukrainian aircraft engine factory, Motor Sich, is a contender. This huge Soviet-era company is one of the advanced military aircraft engine manufacturers in the world. Thirty-five per cent of the company's $450 million in sales in 2018 went to China, making it the company's biggest destination for its aircraft engines. China is interested in Ukrainian technology beyond Motor Sich. They are hiring Ukrainian engineers also in the missile and aircraft-building sectors and taking them to China. There are still question marks on the airborne radar and stealth technologies. However, significant research is going on in state-owned universities and research institutes. Also, China has been pushing hard for arms exports, albeit mostly to relatively poor developing countries by offering price concessions, and attaching some political strings.

AVIATION INDUSTRY CORPORATION OF CHINA

The Aviation Industry Corporation of China Ltd (AVIC) is a state owned aerospace and defence conglomerate ranked 151st in the Fortune Global 500 list.[6] It has over 100 subsidiaries, 27 listed companies, and 446,613 employees across the globe. Established on 1 April 1951 during the Korean War as the Aviation Industry Administration Commission, after many systemic reforms, it got its current designation. The AVIC was founded in November 2008 through the restructuring and consolidation of China Aviation Industry Corporations I and II. Centred on aviation, their business units cover defence, transport aircraft, helicopters, avionics and systems, general aviation, research and development, flight testing, trade and logistics, assets management, finance services, engineering and construction, automobiles and more. AVIC purchased American aircraft engine manufacturer Continental Motors in 2010, American aircraft manufacturer Cirrus in 2011, and American specialized parts supplier Align Aerospace in 2015. The major focus of AVIC is to efficiently develop indigenous military technologies, and to eventually compete with Airbus and Boeing in the civilian airline industry.

Chengdu Aerospace Corporation

The Chengdu Aerospace Corporation[7] (CAC) is a subsidiary of AVIC. It was founded in 1958 in Chengdu, Sichuan province. It designed and now produces the J-10 light-weight multi-role fighter and the J-20 5th generation jet fighter, both of which are considered the most advanced platforms in China's inventory. They also produce the CAC/PAC JF-17 Thunder light-weight multi-role fighter in cooperation with Pakistan. Chengdu Aerospace is credited with being the second country in the world and first in Asia to possess fifth-generation and stealth technology. It employs 20,000 workers. CAC earlier produced the FT-5 trainer, J-7, and the McDonnell Douglas MD-80 parts. CAC later began producing Airbus

A320 and Boeing 757 components, such as vertical and horizontal tails. The first flight of the Chengdu J-10 fighter in 1998 gave the company ability to produce 3rd generation aircraft. The plant also produced fuel tanks for Dassault Falcon 2000EX. The Engine Division of CAC produced the WP6 turbojet LM WP13 turbojet (a Chinese version of Tumansky R-13 engine), and components for Pratt & Whitney JT8D turbofan engines. CAC also manufactures the top end Pterodactyl 1 and Xianglong UAVs.[8]

Xi'an Aircraft Industrial Corporation

The Xi'an Aircraft Industrial Corporation[9] was established at Xi'an in 1958. It employs 20,000 persons. Its main products are the MA-60, MA 600 and under development M700 turbo-prop airliners. It also built the JH-7 Flying Leopard twin-engine fighter-bomber, and the H-6 twin engine bomber (a Chinese upgraded variant of the Tu-16 Badger). Currently, the H-20 strategic bomber is under development. It also manufactures the Y-7H trainer and the wings and fuselage of the ARJ21. Twin engine turboprop transports Y-7 and Y-14 and four engine Y-20 are also manufactured here. It specialised in transport aircraft and built many variants of Y-8. Y-9 was a stretched variant of the Y-8 with greater payload and was China's attempt to build a C 130 J class transport aircraft.

Changhe Aircraft Industries Corporation

The Changhe Aircraft Industries Corporation[10] (CAIC) is a helicopter manufacturer based in the city of Jingdezhen in Jiangxi province. Changhe employs 4,300 employees in two production facilities. It has a joint venture with Augusta Helicopters and a working relationship with Sikorsky Aircraft Corporation. A subsidiary factory is a major automobile company in China. Established in 1969, it manufactures WZ-10 attack helicopters, Z-8 heavy transport helicopters, and a few light utility helicopters. It also makes tail rotor pylons for the Sikorsky S-92 and fuselage for the Sikorsky S-76.

Hongdu Aviation Industry Group

The Hongdu Aviation Industry Group[11] Ltd. is based at Nanchang, and was established in 1951 and has 20,000 employees. It built the Q-5 'Fantan' (exported under the designation A-5)—single-seat dual-engine supersonic attack aircraft based on the MiG-19. Q-6 was a variable sweep-winged attacker, similar to MiG-23BN but was cancelled later. It also built the prototype of the J-12, a 1970 Chinese lightweight supersonic fighter that was later abandoned. The group builds most of the Chinese trainers including JL-8 and L-15. It also built the multi-use agriculture and forest aircraft N-5 and many multi-purpose helicopters of the MD series.

Guizhou Aircraft Industry Corporation

The Guizhou Aircraft Industry Corporation[12] products include trainers, turbojets, UAVs, missiles, and launchers. The JL-9 is a trainer based on MiG-21U. The JJ-7 is an upgraded variant of the same. They also build the WZ-2000, Soar Dragon, and Harrier Hawk UAVs. Guizhou is also the plant to manufacture WS-13 jet engines used in JF-17 and J-31 fighters.

Harbin Aircraft Industry Group

The Harbin Aircraft Industry Group,[13] was founded in 1952 at Harbin. It has 6,700 employees, and was set up to manufacture domestic civil planes. Initially the plant made the Russian helicopter Mi-4, and light bomber H-5, a copy of IL-28. Later it started making indigenously designed Y-11 light twin-engine utility aircraft and Y-12 utility STOL transport variants of Y-11. Harbin has also been producing Legacy 650 and ERJ 145 regional jets in a joint venture with Embraer. Also manufactured here is the BZK-005 high-altitude, long-range UAV designed by Beijing University of Aeronautics & Astronautics for use by the Chinese Navy.

Shanghai Aircraft Manufacturing Company

The Shanghai Aircraft Manufacturing Company,[14] is into manufacturing aircraft, parts and components, repair and overhaul, and many non-aerospace products. It now part of the Commercial Aircraft Corporation of China (COMAC), which was established in 2008. It produced ARJ21, Shanghai Y-10, MD-82, MD-83, and MD-90 jet liners. It also manufactures the Airbus single aisle family cargo door frame, Boeing 737 tail section assembly, and Boeing 777 vertical stabilizers. It also makes the C919, a locally developed narrow-body twinjet airliner. The C919 rolled out on 2 November 2015 and the aircraft's maiden flight was on 5 May 2017.[8] The fourth prototype made its maiden flight in August 2019. Its first commercial deliveries are expected in 2021. The aircraft is to be powered by either CFM International LEAP or ACAE CJ-1000 turbofan engines and will be able to carry 156 to 168 passengers. It is intended to compete primarily with the Boeing 737 MAX and Airbus A320neo. As of

31 August 2018, COMAC reportedly had 1,008 commitments including 305 firm orders, mostly from Chinese leasing companies or airlines. The Shanghai Vinage Airship Manufacturing Company has been making non-rigid airships. The Shenyang Sailplane factory makes sailplanes and training gliders including motorized ones. China has also been making light electric aircraft and motor gliders.

Aircraft Nomenclature

Chinese aircraft are prefixed with alphabets donating the type of aircraft in Chinese language (Figure 1).

Prefix Alphabet	*Chinese Designation*	*Aircraft Type*
H	Hongzhaji	Bomber
J	Jianjiji	Fighter
JH	Jianjiji Hongzhaji	Fighter Bomber
JJ	Jianjiji Jiaolianji	Fighter Trainer
JZ	Jianjiji Zhenchaji	Reconnaissance Fighter
Q	Qiangjiji	Ground Attack Aircraft
JL	Jiaolianji	Trainer
Y	Yunshuji	Transport Aircraft
Z	Zhishengji	Helicopter
U	You	Tanker
CJ	Chuji Jiaolianji	Primary Trainer

Figure 1: Chinese Military Aircraft Designation: Type Prefix

Bombers

Among the bombers manufactured by China was the H-5, a licence-built version of the Russian Ilyushin Il-28, and it has since been retired. The H-6, a licence-built version of the Russian Tupolev Tu-16 is produced at Xian. It had many variants and the aircraft is still flying with PLAAF. H-6K is a cruise missile carrying modernized bomber variant. The latest version is the H-6N, a heavily redesigned version capable as FRA and carrying ALCM. H-20 is a Chinese stealth bomber under development. H-20 could have a low-probability-of-intercept AESA radar, and fuse that information to other firing platforms. The H-20 could also be used for electronic warfare or to deploy specialized directed energy.[15] The JH-7 Flying Leopard is a two-seat, twin-engine fighter bomber in service with the PLAAF and PLANAF. The latest variant JH-7E was shown at the 2018 Zhuhai Air Show.

Fighter Aircraft

The Chinese started their fighter aircraft production with the licensed production of the Russian Mikoyan MiG-9 and MiG-15 (J-2). Subsequently they built many Chinese variants of Russian aircraft, J-5 (MiG-17), J-6 (MiG-19), and J-7 (MiG-21). They developed FC-1(JF-17) multirole light fighters jointly with Pakistan. Q-5 was a ground attack aircraft based on the MiG-19. AVIC built Chinese indigenous fighters J-8 and J-10 with reverse engineering or cut-and-paste designs from Russian aircraft. J-11 was a Su-27 variant. J-15 is a Chinese multirole 4th generation naval aircraft, and J-16 is a 4th generation strike aircraft. J-20 is a 5th generation Stealth fighter and J-31 a 5th generation stealth fighter still under development.

J-31 Stealth Fighter.
Image Source: defensetalk.com

Transport Aircraft

China built many transport and passenger/communication duty aircraft. They initially licence-built Yak-12 aircraft and Y-6 (Il-14). Thereafter they made many on their own by reverse engineering. Y-5 was the copy of Russian An-2, Y-7 (An-24 copy), Y-8 (An-12 copy). Y-9 was a multi-purpose variant of Y-8. Y-10 and Y-11 were small utility aircraft. Y-20 is a large multi-purpose transport aircraft (66 tonne payload) which first flew on 26 January 2013 and was inducted into service in 2016.[16] Y-30 is a midsized, four-engine, turboprop military transport aircraft under development, with a planned load of 30 tons.[17] China is also building many mid-sized turboprop and jet airliners, and business jets. They are working on developing larger airliners with up to 400 seat C-939.

Y-20 Transport Aircraft.
Image Source: Wikipedia

Helicopters

Chinese helicopter production began with Z-5, which was based on a Soviet supplied blueprint and was a copy of Mil Mi-4. Z-6 was a turbo shaft engine variant of Z-5. Z-8 was a licence built variant of Aerospatiale SA 321 Super Frelon, and Z-9 a licence built variant of Eurocopter Dauphin. Z-9WA was a utility-cum-reconnaissance variant. WZ-10 is a Chinese attack helicopter designed primarily for anti-tank warfare missions but has secondary air-to-air capability as well. Initiated by chief designer Wu Ximing, the project was co-designed by Kamov Design Bureau of Russia and the 602nd Aircraft Design Institute under a contract with the Chinese government. The helicopter is being built by CAIC. Z-11 is a licence built Eurocopter AS350, a single engine light utility helicopter. Z-11WB is the reconnaissance variant. EC-120 is a light utility helicopter through a joint venture with Eurocopter. Z-12/15 medium transport helicopter of 5/6-tonne class for a range of military and civil requirements built with Eurocopter support. Z-18 is a single-rotor helicopter with tail rotor and a non-retractable landing gear, based on the Avicopter AC313. It can carry 27 troops or five tonnes (11,000 lb) of cargo. WZ-19, is a Chinese 4,250 kg, reconnaissance/attack helicopter variant of Z-9, in service since 2012. China also operates the Sikorsky S70 medium utility helicopter. The Z-20 is a medium lift, 10 ton class helicopter produced in China by Harbin Aircraft Industries Group. It first flew on 23 December 2013. The Z-20F[18] is expected to perform anti-submarine warfare, search and rescue missions, and other shipboard operations on the Chinese Type 055 cruisers and the Chinese aircraft carrier Liaoning. It is thought to be comparable to the U.S. Sikorsky UH-60 Black Hawk helicopter, the civilian S-70 variant of which has been used by the PLA since 1984. Some sources suggest that the Z-20 is a copy of the Black Hawk and link the design to the Black Hawk that was abandoned by U.S. Special Forces in Pakistan during the operation to kill Osama bin Laden on 1 May 2011. China also makes many civil helicopters.

Trainers

China began making basic propeller trainers with the CJ-5 and CJ-6. They also made trainers for fighter aircraft like J-5, J-6, and J-7. JL-8 was the basic jet trainer-cum-attack aircraft developed with Pakistan. JL-9 was the upgraded version of the J-7 trainer. Xian made a specialized version of the Chinese Y-7, HYJ-7 multi-purpose trainer designed to provide training to PLAAF pilots before transferring them to bomber or transport units. The same aircraft is used to train pilots and crews of the H-6 long range bombers. The Hongdu-Yakovlev CJ-7 (L-7) is a two-seat piston engine trainer aircraft jointly developed with Russia from Yak-152.[19] The Hongdu L-15 Falcon[20] is a supersonic advanced training and light combat aircraft build for PLAAF and PLANAF as a lead-in fighter trainer (LIFT). The aircraft resembles a shortened Super Hornet or F-16. It uses two Ukrainian AL -222 turbofan engines, and has a glass cockpit and hands-on-throttle-and-stick (HOTAS) controls which expose trainees to typical 4th generation fighters.

MILITARY AVIATION: NEW DEVELOPMENTS

China's military aviation industry has advanced at an impressive rate over the past decade. Aircraft manufacturing has switched from poor middle quality reverse-engineered copies to conceptually good home-grown products. Chinese military aircraft are not the best in their class, and the Chinese are conscious of that. They have come a long way in flight performance, though they are still catching up, rather than leading with new designs. However, they are good enough considering PLA's military doctrine and mission requirements. China is clear that they have no alternative but to research and develop their own. Chinese systems can be classified as high average in capability, but average in reliability.

Current Developments: PL 15 Missile

The J-11B fighter jet became the latest user of China's self-developed world-class PL-15 air-to-air missile.[21] The J-20 stealth fighter jet had carried the same type of missile and was displayed at the Chinese Air Show 2018 in Zhuhai. Wei Dongxu, a Beijing-based military analyst, compared the

PL-15 with the USA's latest AIM-120 missile.[22] The U.S. media outlet, *The National Interest*, said the PL-15's effective range could be much higher than the AIM-120's 180 kilometres, but that could be an exaggeration, according to Wei. The PL-15 is also equipped with an active electronically scanned array (AESA) radar, which makes evasion very difficult. Wei said the technology of the PL-15 has matured, so it can now be put on a variety of platforms. By equipping the missile, the J-11B's aerial combat capability can be greatly boosted, the report said. Dubbed by Chinese military observers as the "aerial trident", the J-20, J-16 and J-10C fighter jets can carry the PL-15 missiles.

Advances in JF-17

The Pakistan Air Force (PAF) has over a 100 JF-17 'Thunder' aircraft. The newest variant of the JF-17 fighter, Block III, made its maiden test flight in December 2019. The Block III features a new and larger holographic wide-angle head-up display (HUD) and integrated cockpit display similar to the one used by the J-20.[23] The J-20 is China's first stealth fighter. The JF-17 Block III also features a missile approach warning system used on the J-20. The nose cone of the new JF-17 variant is also longer, making it likely that the fighter has incorporated an AESA radar. The variant appeared to have radar-warning receivers aft of the aircraft intakes and on the tail. The upgrades would improve the JF-17's capability to use both short- and long-range air-to-air missiles. Chinese analysts claimed the JF-17 Block III would be able to match an "improved F-16". Nearly 58 per cent of the aircraft's parts are manufactured in Pakistan. These include the "wings, horizontal tail, vertical tail, and forward fuselage", with the remainder coming from China. Yang Wei, a Chinese legislator and chief designer of the China-Pakistan co-developed fighter jet aims to enhance the jet's informatised warfare capability and weapons.[24] The JF-17 uses a Russian-supplied engine and is equipped primarily with Chinese weapons. In terms of numbers, it is the workhorse of the Pakistan Air Force. The chief of the Pakistan Aeronautical Complex said the "Production of subassemblies has already started for the first two 50 Block 3 aircraft, to be assembled next year, and will be followed by another 12 in 2021, 2022, 2023, and 2024." It was contending with India's Tejas and South Korea's FA-50 in Malaysia's new fighter jet purchase plan. Myanmar and Nigeria have reportedly purchased the Chinese-Pakistani warplane.

J-20: Further Developments

China has plans to develop its most advanced stealth fighter, the Jet J-20 into a bomber, electric warfare (EW) aircraft and a carrier-based variant. Reports suggested a two-seat version of the warplane is under development. All current stealth fighter jets feature single-seat, so the potential J-20 variant might become the first two-seat stealth fighter jet in the world.[25] On a highly digitalized future battlefield, large amounts of information can easily overflow the entire control panel of an aircraft. Having a second pilot and a second panel sharing part of the work will be advantageous, the Chinese feel. The J-20 is being serialized and its combat capability constantly upgraded as it is by design highly customizable. Song Zhongping, a military expert and TV commentator, says, "Outfitting the warplane with a second seat allows it to play multiple roles in addition to winning air superiority." [26] The two-seat version can be further developed into a tactical bomber or EW aircraft, he added. Although the FC-31, another Chinese stealth fighter jet, is widely expected to become China's next generation carrier-borne fighter jet, Song said the J-20 can also be modified to fulfil the role. An upgraded J-20 will have improved avionics and fire control systems, more powerful engines and more weapons payload.

Advances in J-16

China's multi-role fighter jet, the J-16, is now covered by a coating that can provide near stealth capability and the jet is now confirmed to be able to carry all types of air-to-surface weapons in precision strikes.[27] Brigade commander Jiang Jiaji, a People's Liberation Army Air Force (PLAAF) pilot, told CCTV that the silver-gray paint covering the J-16 is a kind of cloaking coating that gives the warplane a certain stealth capability, making it nearly invisible to the naked eye and electromagnetic devices. The jet's camouflaged colour scheme makes the aircraft blend with the sky and sea. Jiang also revealed that all types of Chinese air-to-surface weapons currently in service can be installed on the J-16. The J-16 is reportedly capable of carrying at least eight tons of weapons. Although the J-20 is China's more advanced fighter jet, the PLAAF still needs the J-16 as the two types of fighter jets can complement each other.

J-10B Upgrade

The J-10Bs are being equipped with an engine capable of thrust vectoring control allowing extreme manoeuvrability. The manufacturer of the J-10B, AVIC, said in a statement at the Zhuhai 2018 Air Show that the thrust vectoring control technology used on the aircraft was an independently developed innovation, making China one of the few countries in the world to master this key technology.[28] The engine for the aircraft was widely regarded as a weakness in China's military development, but they seem to be catching up, and the State-owned Aero Engine Corporation of China is continuing to work on the engines to provide more thrust and prolong the life span. Despite impressive displays at the air show, China is still 20 to 30 years behind the USA in the overall design of the engines, experts said. The J-20's chief designer, Yang Wei, said in a China Central Television program that the thrust vectoring control engine used on the J-10B can be installed on any fighter jet any time, if needed, including the J-20."[29] Much like the U.S. Air Force with its mix of stealth and non-stealth fighters, the PLAAF is developing a two-tier fighter fleet. Alongside a handful of radar-evading J-20s, Beijing is acquiring hundreds of more conventional J-10s. The single-engine, single-seat J-10, first flew in 1998 and entered front-line service in 2003. Featuring a tail-less delta wing and canards, the 51-feet-long J-10 is externally similar to the defunct Israeli Lavi fighter. The 2018 edition of the Pentagon's annual report on Chinese military capabilities describes the latest J-10C variant as an "advanced fourth generation fighters armed with the latest weapons."[30] As of late 2017, the Chinese air force possessed around 260 J-10s, according to Flight Global's annual survey of world air arms.[31] J-10s account for 15 per cent of Chinese combat aircraft and nearly half of the roughly 600 Chinese warplanes that, in 2018, the U.S. Defence Department considered modern. The new J-10C model, which reportedly entered front-line service in 2018, has a new engine inlet that apparently reduces the plane's radar signature. It may also have an electronically-scanned-array radar. In 2018 the U.S. military possessed no fewer than 2,800 fighters including more than 900 F-16s and hundreds of F-22 and F-35 stealth fighters.

Pakistan has shown interest in China's J-10 fighter jets and Pakistan Chief of Army Staff, Qamar Javed Bajwa, sat in a J-10C during the "Shaheen-VII" China-Pakistan joint air exercise at the end of 2018, However, a final decision is still to be taken. Laos is a contender for the J-10C. Bangladesh, in 2018, sent a delegation to conduct a field survey of the performance of the J-10C. If and when this happens, the orders are likely to be small. Europe's Typhoon and Rafale fighters are considered expensive. In this context, the Chinese feel that the J-10C fighter jets could be attractive.

Next Generation Fighter

After the successful J-20 fighter, China does not want to fall behind in the global race toward sixth-generation fighter jets and is expected to build its own next-generation fighter jets by 2035, said Wang Haifeng, a chief architect at Chengdu Aircraft Research and Design Institute, who also participated in the development of the J-20 and J-10.[32] Some of the new features of a sixth-generation fighter jet would include the ability to command drones, artificial intelligence and even higher stealth capability through aerodynamic design, the periodical reported. New technologies such as laser, adaptive engines, hypersonic weapons, and swarm warfare might also be part on the new aircraft, Wang said that China will choose some of these features and add others that best suit its needs. France and Germany announced that they will jointly build a next-generation combat jet system, which is expected to be operational by 2040. The UK unveiled its sixth-generation fighter jet development program named 'Tempest' in July 2018, and will invite India to join its co-development aircraft program, as reported in the media. Other countries including the USA, Russia, and Japan are also reportedly developing their own sixth-generation fighter jets. Although they remain in the concept stage, the new fighters are likely to emerge in the 2030s or 2040s. China has yet to officially reveal a plan for its next-generation fighter jet, which hardly comes as a surprise as the country seldom announces any in-development weaponry, but it may have already started related research and development. China's believes in having one generation in service, a new one in development, and a next-generation under study. Now that the J-20 is already in service, the development for a new aircraft is also underway. The generational standards for fighter jets have been defined mainly by Western countries but not future standards, said J-20's chief designer, Yang Wei, noting that China will design very different aircraft in the future through true innovation.[33] China has also constructed a 6,620-tonne, 17,000-cubic-metre FL-62 continuous transonic wind tunnel that will be critical in "shaping China's future fighter jet," said a statement released by the Aviation Industry Corporation of China in September 2018.

Y-20, The New 'Heavy'

"The Y-20 can serve as a general platform from which a variety of variants can be derived," Tang Changhong, a Chinese political advisor and chief designer of the Y-20, said at a press conference on 12 March 2019.[34] Tang's remarks came amid reports and predictions made by military experts since 2018 that China was developing Y-20 variants, including an aerial tanker and an early warning aircraft. Feng Wei, a Chinese legislator and Y-20 pilot, said, "it will be refreshing to see the Y-20 this year, and people won't be disappointed". With a take-off weight of 200 tons, the Y-20 can stay aloft for extended periods, making it a great basic platform for tanker and early warning aircraft. Possible variants may also include a mobile hospital and an electronic warfare aircraft. The Y-20 might also replace its current Russian engines with domestically made WS-20 engines not earlier than 2024. "The Y-20 has now entered the formal mass production stage, and intensive regular training with the military is going according to plan," Tang said.

China is in urgent need of an aerial refuelling tanker that has a larger fuel capacity than the HU-6, a tanker developed from the H-6 bomber, for its air force to become a strategic one. Although China also operates a few Russian II-78 tankers, which are much larger than the HU-6, Russia was reluctant to sell more at a reasonable price, leading China to decide to develop its own large tanker. The Y-20 is of similar size to the Russian Il-76 transport aircraft, on which the Il-78 is based. While the Il-76 carries 44 tons of cargo, the Y-20 carries up to 66 tons. The maximum speed and range are somewhat comparable. The Y-20 began service in the PLAAF in 2016, and over 20 have been built by end 2019.

Heavy Helicopters

The 40-tonne class heavy helicopter, jointly developed by China and Russia, is expected to be delivered by 2032.[35] "Russia is more experienced in the transmission system when it comes to 40-tonne class helicopters, as Russia's Mi-26 is of the 56-tonne class. "Our goal in the cooperation is to learn from Russia's strong points and close the gap," said Wu Ximing, chief designer of helicopters for the AVIC. Wu accepted that China lacks experience in technologies related to the transmission system. After four years of negotiations, Russia's state corporation Rostec has agreed to sign the contract. Under the contract, at least 200 heavy helicopters will be built in China. China would be responsible for the helicopter's design and production and Russia would be acting as a technical partner. The helicopter, dubbed Advanced Heavy Lift, would carry a payload of 15 tons, and have a range of 630 kilometres and a top speed of 300 kilometres an hour. China will have a complete helicopter family from 500-kilogram class to 40-tonne class, Wu said.

Unmanned Aerial Systems

China's domestically developed, made-for-export Wing Loong series of drones have fired more than 3,000 rounds of live munitions with an overall accuracy higher than 90 per cent, the Chinese media reported.[36] Wing Loong is an armed reconnaissance drone capable of delivering precision strikes using air-to-ground missiles, often likened to the US's MQ-1 Predator and MQ-9 Reaper drones. As of December 2018, 100 Wing Loong series drones were delivered for export.[37]

A video featuring China's flying saucer-like stealth drone, the Sky Hawk, was shown for the first time on China Central Television (CCTV) in January 2019.[38] It has been developed by China Aerospace Science and Industry Corporation. Featuring a "flying wing" aerodynamic design similar to the US B-2 stealth bomber, the Sky Hawk is a high-altitude, long-range and high-speed unmanned aerial vehicle capable of conducting reconnaissance and patrol missions in hostile environments. Another Chinese stealth drone with a 'flying wing' design, the CH-7, developed by China Aerospace Science and Technology Corporation, was also on display at the last Zhuhai air show. Its 22-metre wingspan makes it significantly larger than the Sky Hawk, providing another choice for domestic and international users. The USA has developed the X-47B stealth drone and run tests on aircraft carriers. The Sky Hawk will also operate on China's future aircraft carriers that will use electromagnetic catapults. China's new strategic bomber, the H-20, is expected to also use a 'flying wing' aerodynamic design to gain stealth capability and other benefits.

Capable of delivering precision bombardment, the weirdly shaped Chinese helicopter drone, Blowfish A2, sparked interest from the militaries of many countries following its flight demonstration at the 15th Langkawi International Maritime and Aerospace Exhibition (LIMA) in Malaysia held in March 2019.[39] The 1.87-metre long, 0.62-metre tall helicopter drone has a maximum take-off weight of 38 kilograms and is capable of carrying a 12-kilogram payload, says Guangdong-based Zhuhai Ziyan UAV Company, the manufacturer of the drone. The Blowfish A2 can carry radar, jamming devices, guns, or bombs under its spine, and has a maximum speed of 130 kilometres an hour. A video shows the drone dropping four bombs, which detonated some metres above the target, as the guided explosions accurately scorched a wide area. The drone is available for export and multiple countries have reportedly shown interest. Combat-ready helicopter drones made by Ziyan now operate in four countries in the Middle East, Southeast Asia, and Africa, the representative said, without naming the clients. AVIC is also developing helicopter drones including the missile-carrying AV500.

AVIC expects to produce 100 high-end drones per year by 2025. AVIC also announced it has established a new subsidiary, AVIC (Chengdu) Unmanned Aerial Vehicle System Company, focusing on the drone business, which Chinese military experts said will help AVIC become even more competitive on the international market. AVIC now offers products like the Wing Loong armed reconnaissance drone, the Cloud Shadow high-altitude drone, the AV500 unmanned helicopter and the Yaoying remote sensing drone. The establishment of the subsidiary shows that AVIC takes the drone business very seriously, and the move is also motivated by the strong demand from the international market. "The US General Atomics is the market leader in the international drone industry. I hope that our new company can surpass it," Lai Zhiyong, an employee at AVIC (Chengdu) Unmanned Aerial Vehicle System Co. Ltd, says.[40] Li said that the reason behind the success of the Chinese drone on the international market is its high quality and low pricing, and China continues to make rapid progress in related technologies. Countries like Egypt, Indonesia, and Serbia are operating the Wing Loong I drone. The Wing Loong I-D, the new Wing Loong series drone, uses an all-composite material structure and made its maiden flight in January 2019.

Electronic Warfare

China has also reportedly developed a new type of electronic warfare aircraft with extra antenna installations. The aircraft appears to have been developed from the Y-9, says Wei Dongxu, a Beijing-based military analyst.[41] The Y-9, a medium-sized tactical transport plane has been modified, to include use as an early warning aircraft, reconnaissance plane and anti-submarine aircraft. The new variant visibly has a hemispheric radar dome under its chin, two large antennas on each side of the plane, an antenna on each side of the tailfin and an electronic warfare pod on top of the tailfin. It could effectively monitor enemy radio communication and intercept their radar signals. It can also deliver electronic suppression, supporting China's aerial strike formations by jamming and paralyzing hostile air defence systems. However, more aggressive missions would be supported by fighter class EW aircraft. The aircraft is likely to be designated GX-11.

Aircraft Carriers

A Type-002 aircraft carrier has reportedly completed several sea trials. Resembling the 'Liaoning', the new aircraft carrier is fitted with a ski-jump assisted short take-off. The Type-002 aircraft carrier has reportedly been fitted with a new generation of phased array radar system, optimized air traffic command room, and flight deck. Its carrier aircraft will also be a new generation of ship-borne J-15 fighter jets equipped with phased array radar. That translates into considerably enhanced combat capability.

Having already designed China's current aircraft carrier-borne fighter jet, the J-15, Shenyang Aircraft Design Institute is developing a new carrier-based warplane based on the FC-31. The FC-31 is a fifth generation medium-sized stealth fighter jet originally intended for export. The FC-31 made its public debut flight at Air Show China 2014 in Zhuhai, but went relatively quiet after that. Multiple changes and upgrades are being made to the FC-31 allowing it to be used on an aircraft carrier. China's third aircraft carrier is widely expected to be equipped with an electromagnetic catapult and likely to house the stealth fighter jet.

Su-30 MK2: Anti-Shipping Role

China unveiled a set of photos of the Chinese People's Liberation Army (PLA) Navy's Su-30MKK fighter jets on 06 January 2019.[42] These photos 'inadvertently' revealed a Su-30 MK2 fighter preparing to take off mounted with a Chinese PL-12 air-to-air missile. This indicates that China has already been able to modify the Su-30MKK's fire control system to give it the capability to use China's indigenous weapons. The next would be the Su-30 MK2 fighter jet carrying YJ-12 and YJ-18 anti-ship cruise missiles. It has enhanced the structure of its total payload to 12 tons. The Su-30MKK also has increased its maximum take-off weight. Its maximum range is almost 4,000 kilometres, and therefore it can patrol the entire South China Sea with the support of tanker aircraft. The Su-30MKK fighter jet, H-6 Strategic Bomber and JH-7 Fighter-Bomber together shoulder the responsibility of China's long-range anti-ship attack task. The Su-30MKK fighter jet has three 2-tonne heavy hanging points and therefore can mount three YJ-12 missiles, supersonic anti-shipping Cruise missiles.

PLAN's New VTOL Drone

The Chinese People's Liberation Army Navy (PLAN) deployed a new vertical take-off and landing (VTOL) fixed-wing drone on a guided missile destroyer in an exercise in the South China Sea in late February 2019 for the first time.[43] The drone took off from the helicopter deck of the Lanzhou, a Type 052C destroyer. The drone has a triple-fuselage design. The left and right fuselages each have four propellers: two on top, two at the bottom. A larger propeller is installed at the rear of the aircraft. "The eight smaller propellers can provide the lift needed for VTOL, and the larger propeller provides thrust", it was reported. The new drone has combined the advantage of a rotorcraft and a fixed-wing aircraft. The drone appears to have a wing span of about four metres, allowing it to be stored in a helicopter hangar. The drone is likely to carry out reconnaissance and search missions at longer ranges for destroyers and frigates. The drone could also guide artillery fire from warships and conduct damage assessment during an amphibious landing operation.

Multiple Terahertz Radiation Radar

Chinese arms companies recently made multiple terahertz radiation radar systems with a technology seen by experts as an efficient air-to-ground reconnaissance tool and a potential counter to stealth aircraft. The prototype radar was successfully developed by the China Electronics Technology Group Corporation (CETC), and a second-generation prototype is already in development.[44] The CCTV report said that terahertz radiation has wavelengths between those of infrared rays and microwaves, a wide spectrum that would render current stealth technologies obsolete, making the radar able to detect stealth aircraft. Stealth aircraft usually use composite materials and radar wave-absorbing coatings, so normal radars cannot effectively detect them. Terahertz radiation, on the other hand, could penetrate those materials and expose metallic parts within the aircraft, according to Wei Dongxu. Experts said the terahertz radiation decays very fast in the air, meaning, the effective range of the radar is likely low and not sufficient to detect an advanced stealth fighter jet in time before it launches attacks from beyond visual range. The China Aerospace Science and Industry Corporation (CASIC) had earlier developed China's first terahertz radiation video synthetic aperture radar, the Beijing-based newspaper, *Science and Technology Daily*, reported in December 2018. The CASIC radar can see through complicated environments like smoke, smog, and dim lights, and can efficiently detect ground infantry targets in camouflage and disguise, and deliver precision strikes.

Early Warning Radar Technology

China is a major arms exporter, but its image in the international weapons market has long been linked to old, second-tier products sold at relatively low prices. Domestic defence technology companies have been sparing no effort over the past several years to improve their reputation by promoting modern, advanced products featuring the latest technology. One recent effort is an airborne early-warning and control aircraft. Hu Mingchun, head of the Nanjing Research Institute of Electronics Technology in Jiangsu province, said there are only a handful of nations including China, the USA, and Israel that can design, build, and export such cutting-edge hardware as early-warning planes. The KLC-7 Silk Road Eye developed by his institute was a generation ahead of its rivals in the global market he claimed.[45] According to Hu, the KLC-7 integrates a mechanical scanning system with active electronically scanned arrays and features the latest digital technology and processing capacity. The system boasts better anti-jamming functions and a longer detection range. The electronics institute in Nanjing, which is part of the State-owned defence giant, the China Electronics Technology Group Corp, is the country's top developer of military surveillance radar. Its products have been sold to more than 20 nations in Africa and Asia, it said. It designed and manufactured the radars mounted on the ZDK-03 early-warning aircraft that China exported to Pakistan. In a picture released by the China Electronics Technology Group Corp, the Silk Road Eye appears similar to the ZDK-03, which means it is also mounted on the Y-9 turboprop transport plane built by AVIC.

Aviation Industry Comes of Age

China's defence industry is exploding onto the scene as its top arms makers push past Western powerhouses.[46] Till 2018, not a single Chinese company had even cracked the world's top 100 defence firms, according to a list published annually by *Defence News*. In 2019, six Chinese defence firms were among the world's top 15, with Chinese companies occupying eight of the top 25 spots. AVIC, with its annual defence revenue close to US$ 25 billion, ranks fifth, outpacing US and UK defence giants General Dynamics and BAE Systems. AVIC, the top Chinese company on the list, is trailing closely behind Raytheon and Northrop Grumman, two leading US defence firms. AVIC is responsible for the development of China's fifth-generation J-20 fighter and the new H-20 stealth bomber, among other projects.

AVIC has been developing Wingman drones and an unnamed flying wing stealth drone was showcased at the 2018 Zhuhai Air Show, reported Liu Xuanzun.[47] The USA and Russia already have drones that will accompany manned fighters and China is conscious that they must not get left behind. The USAF is inducting XQ-58 Valkyrie drones to fly in formation with the F-15EX and F-35. Similarly Russia has just released footage of the first flight of the S-70 Okhotnik stealth assault drone to be paired with the Su-57 for joint missions. The AVIC has already developed the Dark Sword stealth drone of the size of a fighter aircraft. The Sky Hawk and Sharp Sword are other stealth drones.

'Made in China 2025' Goals

China wants its commercial aircraft to supply 10 per cent of the domestic market and its jetliners to account for up to 20 per cent of the global market by 2025.[48] When the C919, China's indigenous long-haul airliner, successfully completed its maiden flight in May 2017, officials were quick to announce that the country was edging closer to clinching the "crown jewels" of the modern aircraft manufacturing industry, and also carried the huge weight of a nation's ambition to be a major player in the global aviation industry. China's civil aviation market is expected to overtake the US as the world's largest by civil 2022, and the country is estimated to need over 7,000 planes in the next 20 years. Being able to make its own would not only help disentangle Beijing from the political complications of having to deal with the global manufacturing duopoly of Boeing and Airbus, but would also be a statement that it had the technological prowess to match its economic might. "China is playing the long game, it's not about the 2020s. China is looking at the next 20, 30, or 40 years," says Kevin Michaels, managing director of AeroDynamic Advisory, an aerospace and aviation consultancy. China's aviation industry has received over 800 orders and options to buy the C919 as of June 2019, despite the fact that it is not expected to enter operation until at least 2021 and that only 50 per cent of the components are currently domestically produced. Among the obstacles the country faces in becoming an aviation power are its lack of expertise in avionics, materials technology and aerodynamics, and most crucially, engines. Experts have estimated that China's jet engine technology is about 20-to-30 years behind its competitors. The C919 currently runs on engines from CFM International, a joint venture between GE Aviation of the USA and France's Safran Aircraft Engines. Beijing has invested 100 billion Yuan (US$ 14.4 billion) in 2016 under the MIC2025 plan to establish the Aero Engine Corporation of China (AECC), which will build the CJ-1000A turbofan jet engine to power the C919. The USA has put the brakes on the transfer of critical technologies. In fact, China has spent less than US$ 1 billion on 12 aviation-related acquisitions in the USA over the decade up to 2017, according to estimates by the US think-tank Rand Corporation. Meanwhile COMAC has 16 joint ventures with foreign firms, including GE Aviation, Honeywell, Parker Aerospace, and Liebherr, and is also building a larger twin-aisle plane, the C929, with a Russian company, but it remains an uphill task as a whole in the form of a new-found US hostility to it. COMAC is yet to apply for certification for the C919 from the US Federal Aviation Administration, without which it cannot access the US market.

Further Russian Fighter Offers

Since the Soviet Union's collapse, the China has been a leading client for high-end Russian military hardware. Such imports have played a key role in revolutionizing China's aerial warfare capabilities, in particular during the 1990s with the acquisition of the most capable Russian air defence systems, air-to-air missiles and air superiority fighters available, but by the 2010s the reliance on high-end Russian equipment was reduced considerably. Russia's was unable to market more than two dozen of its latest Su-35 fighter jets in 2015, an offer that was accompanied by generous technology transfers to be accepted. Upon announcing plans for the export of Russia's Su-57 fifth generation air superiority fighter at the 2019 LIMA exhibition in Malaysia, Kladov Viktor, of Rostec, the Russian state conglomerate for arms exports, mentioned the aircraft's potential export destinations.[49] In the next two years, China will make a decision to either procure additional 4th Gen ++ Su-35s, or build the Su-35 within China, alternatively procure Su-57E, he said.

However China's induction of its own fifth generation air superiority fighter, the Chengdu J-20, in 2017, preceded the entry of the Su-57 into service and marked the first aircraft of its generation to complete development outside the USA. China also preceded Russia in its deployment of aircraft with active electronically scanned array radars, which are currently mounted on its J-20 and J-10C fighters, as well as its deployment of next generation air-to-air munitions with the entry into service of the PL-15. While the Su-35 did field a number of technologies that China's own defence sector had yet to develop, including three-dimensional thrust vectoring AL-41F-1S engines, these technologies are expected to have been mastered by Chinese military aviation by 2022 and deployed on the upcoming J-11D. Should the J-11D program succeed, China would have little reason to purchase further Su-35 aircraft or to seek to establish production of the platform domestically, as director Kladov suggested. The Su-57's internal weapons payload is much larger than that of the J-20, deploying eight long range K-77 or R-37 long range air-to-air missiles. That also gives the Su-57 a considerably longer air-to-air engagement range compared to the J-20, with the K-77 missile retaining a range of over 193 kilometres. The Su-57's internal missile bays can also deploy the R-37 with a 400 km range—designed specifically to target tankers, airborne warning and control systems (AWACS), and other vital support systems. The Su-57 also allows it to fire modulated laser beams at seekers of incoming missiles to blind them. These are all capabilities that existing Chinese fighters, including the J-20, lack entirely. The Russians claim that the Su-57 outclasses the J-20 across the spectrum. With large scale production plan of the indigenous J-20, the Su-57 may be ruled out. At best a few Su-57 fighters could be bought for considerable technology transfers.

Critical Technologies China Still Needs

China cannot still make large airplane engines that can efficiently power large commercial aircraft. In order to make those engines, China needs to first make light-weight blades that can keep their strength and shape while holding up to extreme temperatures (nearly 2,000° Fahrenheit) and very high rotation.[50] The engines for commercial planes are required to have a longer life span and endure 3,500 to 5,000 hours of use a year. China cannot even do reverse-engineering because it needs advanced machine tools for turbine blades. The USA and other Western countries won't sell the advanced machine tools to China, and the ones they have purchased aren't precise enough to mass-produce the intricate blades needed for the engines. "China continues to lag behind in avionics for advanced fighter jets and until Chinese companies develop capabilities on par with leading competitors from the USA and Russia, they will likely double down on industrial espionage to steal designs,"[51] said Bonnie Glaser, senior advisor for Asia at the Centre for Strategic and International Studies. "China has developed techniques to target U.S. defence contractors and hack defence industry computers," Glaser added. "Both methods are proving effective in acquiring the most advanced military assets." According to publicly available data about the J-20, its avionics include electronic surveillance measures, infrared search-and-track and electro-optical targeting systems, and third-party sensor data fed to it via data-link to help locate its targets. Experts believe the J-20's avionics is most likely comparable to that installed on the F-22 and F-35 fighters, but of lower quality.

Estimates show that China has derived a cost advantage of just 14 to 20 per cent, which is hardly impressive given that China is the country that has relied most extensively on both industrial and cyber espionage, and that it has benefited massively from the transfer of technology through FDI, mergers and acquisitions, and the purchase of foreign components.[52] Such a cost advantage is even less impressive given that the F-22 is now twelve years old, that the J-20 has significant deficiencies, and that its costs will inevitably increase further as China attempts to fix its problems and improve its performance. The latter point is critical, because "the final 10 per cent striving towards maximum perfection costs 40 per cent of the total expenditure on most projects." The same is true with regard to time. The F-22 became operational in 2005—about twenty years after the project started. Launched in the late 1990s and tested in 2010, the J-20 was officially commissioned in late 2017. Still, it is not yet fully operational and remains inferior to the F-22 on several dimensions: in other words, after more than twenty years, China has not yet closed the gap with the USA.

China's Emerging Aviation Superiority

China's defence industrial base is working to modernize its military to build a world-class force that can fight and win wars, an ambition repeatedly stressed by the Chinese leadership. "The Soviets were never able to match, much less overcome, America's technological superiority. The same may not be true for China," former Deputy US Defence Secretary Robert Work and his colleague Greg Grant wrote in a recent report.[53]

China's economic power makes it highly unlikely that the USA will be able to spend its way to victory in its strategic competition with China, the authors contend. The USA has not faced a competitor with a Gross Domestic Product (GDP) greater than 40 per cent of its own in more than a century. China's GDP is currently around 63 per cent of that of the USA, and China is projected to eventually have the world's largest economy. "China also has the political will and fiscal strength to sustain a steady increase in defence spending during the next decade," the Department of Defence explained in its 2019 report on China's military might, noting that these increases "will help support PLA modernization, develop an integrated military-civilian defence industry, and explore new technologies with defence applications." The Pentagon identified the key elements of China's military modernization as investments in domestic defence, the development of the defence industrial complex, a growing science and technology research and development base, civil-military integration, and the acquisition of foreign technology. " The result of this multifaceted approach to technology acquisition is a PLA on the verge of fielding some of the most modern weapon systems in the world," Lt. Gen. Robert Ashley, the director of the Defence Intelligence Agency, wrote in a letter prefacing a 2019 DIA report on China's military modernization.[54] "In some areas," he added, "it already leads the world."

Options for India

As China has set a target to become a leading super power by the year 2049, it is in competition with U.S.A., the current super power whom it wants to replace. Its size of the economy is growing at a fast rate. It is already ahead of America's GDP in PPP terms and well on track to overtake it even in nominal terms. In the recent October 2019 parade, it displayed its military might. It is trying to persuade nations to accept its new position by enticing them with financial help by using its massive economic clout, to accept its extraterritorial claims willingly or if necessary with its threatening posture and nations, which do not succumb to, pressures, then by being overtly friendly. India falls in the last category after the Doklam standoff and in Galwan, India stuck to its position. Displayed resolve and matched force for force.

'Make in India' defence production has finally started showing initial results. India is way behind, but all national energies need to remain focused on the aim for India to take its place at the global high table.

Aviation technologies are much tougher to master. India needs to take foreign help to master them. It also needs to invest much more on research and development. Also defence production needs to be released from bureaucratic control as has been successfully done for India's space and nuclear programs.

REFERENCES

1. China celebrated 70 years of communist party rule, Zee News, 01 October 2019, https://zeenews.india.com/world/china-unveils-dongfeng-41-missiles-that-can-strike-us-in-30-minutes-2237836.html
2. Wendell Minnick, Russia-China Su-35 Deal Raises Reverse Engineering Issue, Wendell Minnick, *Defence News*, November 20, 2015, https://www.defensenews.com/air/2015/11/20/russia-china-su-35-deal-raises-reverse-engineering-issue/
3. Nick Childs, China's naval shipbuilding: delivering on its ambition in a big way, IISS Military Balance Blog, May 01, 2018, https://www.iiss.org/blogs/military-balance/2018/05/china-naval-shipbuilding
4. Chinese military tests long-range, heavy aerial delivery transport aircraft, *Global Times*, June 23, 2019 http://eng.chinamil.com.cn/view/2019-06/23/content_9536409.htm
5. Kyle Mizokami, The J-10 fighter plane gets thrust vector control, *Popular Mechanics*, Nov 07, 2018, https://www.popularmechanics.com/military/a24742661/chinas-j-10-thrust-vector-control/
6. Aviation Industry Corporation of China, Fortune Global 500, https://fortune.com/global500/2019/aviation-industry-corp-of-china/
7. Wikipedia, Chengdu Aircraft Industry Group, https://en.wikipedia.org/wiki/Chengdu_Aircraft_Industry_Group
8. Chengdu (AVIC) Wing-Loong (Pterodactyl), Military factory, September 11, 2018.
9. Wikipedia, Xi'an Aircraft Industrial Corporation, https://en.wikipedia.org/wiki/Xi%27an_Aircraft_Industrial_Corporation
10. Wikipedia, Changhe Aircraft Industries Corporation, https://en.wikipedia.org/wiki/Changhe_Aircraft_Industries_Corporation
11. Wikipedia, Hongdu, https://en.wikipedia.org/wiki/Hongdu
12. Wikipedia, Guizhou Aircraft Industry Corporation, https://en.wikipedia.org/wiki/Guizhou_Aircraft_Industry_Corporation
13. Wikipedia, Harbin Aircraft Industry Group, https://en.wikipedia.org/wiki/Harbin_Aircraft_Industry_Group
14. Wikipedia, Shanghai Aircraft Manufacturing Company, https://en.wikipedia.org/wiki/Shanghai_Aircraft_Manufacturing_Company
15. Sebastien Roblin, Danger: Should We Fear China's New H-20 Stealth Bomber? *The National Interest*, January 3, 2020, https://nationalinterest.org/blog/buzz/danger-should-we-fear-chinas-new-h-20-stealth-bomber-110681
16. Air Marshal B.K. Pandey, China's Strategic Airlift Capability, No Man's Land July 7, 2016 http://nmlonline.weebly.com/analysis—review/chinas-strategic-airlift-capability
17. Chinese Y-30 Medium Turboprop Military Transport Aircraft, *Chinese Military Review* http://chinesemilitaryreview.blogspot.com/2014/11/chinese-y-30-medium-

turboprop-military.html

18. Mike Yeo, Chinese media reveals specs of Harbin Z-20 helo, *Defence News*, October 14, 2019, https://www.defensenews.com/global/asia-pacific/2019/10/14/chinese-media-reveals-specs-of-harbin-z-20-helo/
19. Wikipedia, Hongdu Yakovlev CJ-7, https://en.wikipedia.org/wiki/Hongdu_Yakovlev_CJ-7
20. Dave Makichuk, China's L-15 Falcon: Cut-rate warfare on a budget, *Asia Times*, January 3, 2020, https://asiatimes.com/2020/01/chinas-l-15-falcon-cut-rate-warfare-on-a-budget/
21. Chen Zhuo, China's most powerful air-to-air missile equipped on warplanes, *Global Times*, March 26, 2019 http://english.pladaily.com.cn/view/2019-03/26/content_9459759.htm
22. Ibid.
23. The Web Desk, China touts use of stealth fighter systems on upgraded jet for Pakistan, *The Week*, January 02, 2020, https://www.theweek.in/news/world/2020/01/02/china-touts-use-of-stealth-fighter-systems-on-upgraded-jet-for-pakistan.html
24. PTI Beijing, China, Pakistan to upgrade their jointly produced fighter jet, *The Hindu* business line, March 12, 2019, https://www.thehindubusinessline.com/news/world/china-pakistan-to-upgrade-their-jointly-produced-fighter-jet-report/article26510862.ece
25. Liu Xuanzun, J-20 variant maybe world's first two-seat stealth fighter, *Global Times*, January 16, 2019 http://www.globaltimes.cn/content/1135957.shtml
26. Ibid.
27. Li Jiayao, J-16 fighter jet stealth-capable, able to carry all types of Chinese air-to-surface weapons, *Global Times*, January 30, 2019 http://eng.chinamil.com.cn/view/2019-01/30/content_9416890.htm
28. William Kucinski, J-10B fighter aircraft debuts Chinese thrust vectoring technology, *SAE International*, November 17, 2018, https://www.sae.org/news/2018/11/j-10b-fighter-aircraft-debuts-chinese-thrust-vectoring-technology-at-2018-china-international-aviation--aerospace-exhibition
29. Huang Panyue, Chinese people amazed by the country's rapid development in fight jets, *Global Times*, November 26, 2018 http://eng.chinamil.com.cn/view/2018-11/26/content_9355534.htm
30. David Axe, Forget Stealth Fighters and Aircraft Carriers: America Should Fear China's J-10 Fighter, *The National Interest*, August 25, 2019, https://nationalinterest.org/blog/buzz/forget-stealth-fighters-and-aircraft-carriers-america-should-fear-chinas-j-10-fighter
31. David Axe, Study This Picture: China's J-10 Fighter Is One Tough Fighter Jet, *The National Interest*, November 10, 2019, https://nationalinterest.org/blog/buzz/study-picture-chinas-j-10-fighter-one-tough-fighter-jet-94751
32. Liu Xuanzun, China eyes building next-generation fighter jets by 2035, *Global Times*, February 11, 2019. http://www.globaltimes.cn/content/1138454.shtml
33. Ibid.
34. Chen Zhuo China's Y-20 large transport plane to spawn several variants: chief designer, *Global Times*, March 12, 2019 http://eng.chinamil.com.cn/view/2019-03/12/content_9448116.htm
35. Liu Xuanzun, 40-ton class heavy helicopter jointly developed by China, Russia to be

delivered by 2032, *Global Times*, March 10, 2019 http://www.globaltimes.cn/content/1141515.shtml

36. Web Desk, Chinese-built armed drones have fired 3,000 rounds of weaponry, *The Week*, April 02, 2019, https://www.theweek.in/news/world/2019/04/02/china-armed-drones-weapons.html
37. Editor Zh, China exports 100 unmanned aircraft Wing Loong, Xinhua, December 25, 2018, http://www.xinhuanet.com/english/2018-12/25/c_137698250.htm
38. Jesse Johnson, China releases first video of a Sky Hawk, its latest stealth drone, in flight, *The Japan Times*, January 6, 2019, https://www.japantimes.co.jp/news/2019/01/06/asia-pacific/china-releases-first-video-sky-hawk-latest-stealth-drone-flight/#.Xm24n3IzbIU
39. Liu Xuanzun, Oddly shaped Chinese combat-ready helicopter drone popular in international market, *Global Times*, April 2, 2019, https://www.globaltimes.cn/content/1144390.shtml
40. Yao Jianing, 100th Wing Long armed reconnaissance drone to be delivered to international client, *Global Times*, December 2, 2018 http://eng.chinamil.com.cn/view/2018-12/27/content_9389500.htm
41. Chen Zhuo, New aircraft 'to gather intel in S. China Sea, *Global Times*, March 07, 2019 http://eng.chinamil.com.cn/view/2019-03/07/content_9444106.htm
42. Huang Panyue, With modified fire control system, China's Su-30 fighter jets might be aircraft carrier's nightmare, China military Online, February 01, 2019 http://eng.chinamil.com.cn/view/2019-02/01/content_9420175.htm
43. Li Jiayao, Rare VTOL drone deployed on PLA destroyer, *Global Times*, February 28, 2019 http://eng.chinamil.com.cn/view/2019-02/28/content_9435911.htm
44. Chen Zhuo, China develops anti-stealth radars, *Global Times*, March 19, 2019, http://eng.chinamil.com.cn/view/2019-03/19/content_9452332.htm
45. Li Jiayao, 'World-class' military plane to be exported, *China Daily*, March 18, 2019 http://eng.chinamil.com.cn/view/2019-03/18/content_9452128.htm
46. Ryan Pickrell, China's defence industry is exploding onto the scene as its top arms makers push past Western powerhouses, *Business Insider*, July 24, 2019, https://www.businessinsider.in/chinas-defense-industry-is-exploding-onto-the-scene-as-its-top-arms-makers-push-past-western-powerhouses/articleshow/70354108.cms
47. Liu Xuanzun, Wingman drones become new trend for fighter jets, *Global Times*, August 11, 2019, http://www.globaltimes.cn/content/1161030.shtml
48. Amanda Lee, China's aviation industry has a steep climb to 'Made in China 2025' goals, *South China Morning Post*, October 29, 2018, https://www.scmp.com/business/article/2170746/chinas-aviation-industry-has-steep-climb-made-china-2025-goals
49. PTI Beijing, China mulls to buy Russia's Su-57 stealth fighter jet, *Economic Times*, April 01, 2019, https://economictimes.indiatimes.com/news/defence/china-mulls-to-buy-russias-su-57-stealth-fighter-jet/articleshow/68666095.cms?from=mdr
50. Qin Chen and Viola Zhou, Inkstone answers: why China can't build a good jet engine and more, Inkstone newsletter, https://www.inkstonenews.com/tech/inkstone-answers-why-china-cannot-build-commercial-aircraft-engines/article/3002672
51. Eugene Gerden, Avionics in China: C919 Development and Future Growth, Avionics International, http://interactive.aviationtoday.com/avionicsmagazine/february-2019/avionics-in-china-c919-development-and-future-growth/
52. Andrea Gilli and Mauro Gilli, Why China Has Not Caught Up Yet, *MIT Press Journal*,

p. 186, https://www.mitpressjournals.org/doi/pdf/10.1162/isec_a_00337

53. Ryan Pickrell, China's defence industry is exploding onto the scene as its top arms makers push past Western powerhouses, *Business Insider*, July 24, 2019, https://www.businessinsider.in/home/chinas-defense-industry-is-exploding-onto-the-scene-as-its-top-arms-makers-push-past-western-powerhouses/articleshow/70354108.cms
54. DIA Chinese Military Power Report, *News USNI*, January 15, 2019, https://news.usni.org/2019/01/15/dia-chinese-military-power-report

5

PLAAF – Modernization:
Work in Progress

"All men can see these tactics whereby I conquer, but what none can see is the strategy out of which victory is evolved"

—Sun Tzu

"Because of its independence of surface limitations and its superior speed, the airplane is the offensive weapon par excellence"

—Giulio Douhet

The PLAAF, the principal instrument of Chinese air power besides the shore based Naval Air Force, was formally organized in late 1949, and specifically incorporated the creation of an air force into the constitution. Communist Chinese air power was thereby mated to the state as a permanent benchmark of China's progress. The fortunes of Communist China as a nation will for ever be mirrored and magnified to the world by the strength or weakness of its air power.

The PLAAF for long has been striving to emulate, and more recently to surpass the USAF. They have set themselves the goal to dominate the air and space power in the international security environment. They have become conscious of the challenges of aerospace power, and they need to maintain high efficiency in all forms of operational engagements. They are driving the establishment of the new international order within a new hierarchy of power structure. They have learnt from the Americans the need to create smart asymmetrical systems. They are using the globalization opportunities and the strong and successful Chinese Diaspora for technological leapfrogging. The current challenges for air and space power and the hybrid nature of warfare, include precision and effect-based operations, high-speed computation and network based operational planning, and flexibility. Air power as a key contributor to any defeat mechanism. Air and space power are the main providers of fire projection on a battlefield and there is a relative shift towards the primacy of air power to deliver kinetic effects.

Seizing the initiative remains the most challenging part of the conventional campaign; gaining air superiority or supremacy to allow freedom of operation to own forces and hindering the offensive of the adversary's force through coercive campaigns. The air force requires capabilities for counter-air, counter-land, strategic attack, command and control, Intelligence Surveillance and Reconnaissance (ISR), Information Operations, mobility and space-based support. Irregular warfare would support air operations. Air and space power also have a great role in counter-insurgency operations. Notwithstanding the dilemma of tactical strikes during counter-insurgency, using combat air power, retaliation, attrition, or interdiction of rebel forces remains a viable option. Urban warfare is also a reality of the times if global influence has to be established. Air and space also greatly support emergency relief operations. Air expeditionary task force ability is important for any big air force.

A key factor to achieve to become a top end air force is a clear vision for the use of air power, availability of funds for defence research and development, acquiring or developing new air and space related technologies. The PLAAF can become a global air force if it is adequately budgeted. Unlike the USAF, the PLAAF currently has low exposure to military campaigns or modern air exercise exposure. Brute numbers or powerful platforms and weapons, though important, can be neutralized through better tactical appreciation or approach.

The PLA is developing its capabilities at a rapid speed, and spending vast sums on breakthrough technologies like artificial intelligence, unmanned systems, hypersonic, and robotics, which could tilt the nature of warfare to its advantage. China's 2019 defence budget[1] of US$ 177.54 billion grew by 7.5 per cent over the previous year, albeit it is still very small vis-a-vis the US defence budget of US$ 750 billion. China's military modernization objectives remain as ambitious as ever. Xi has also begun a

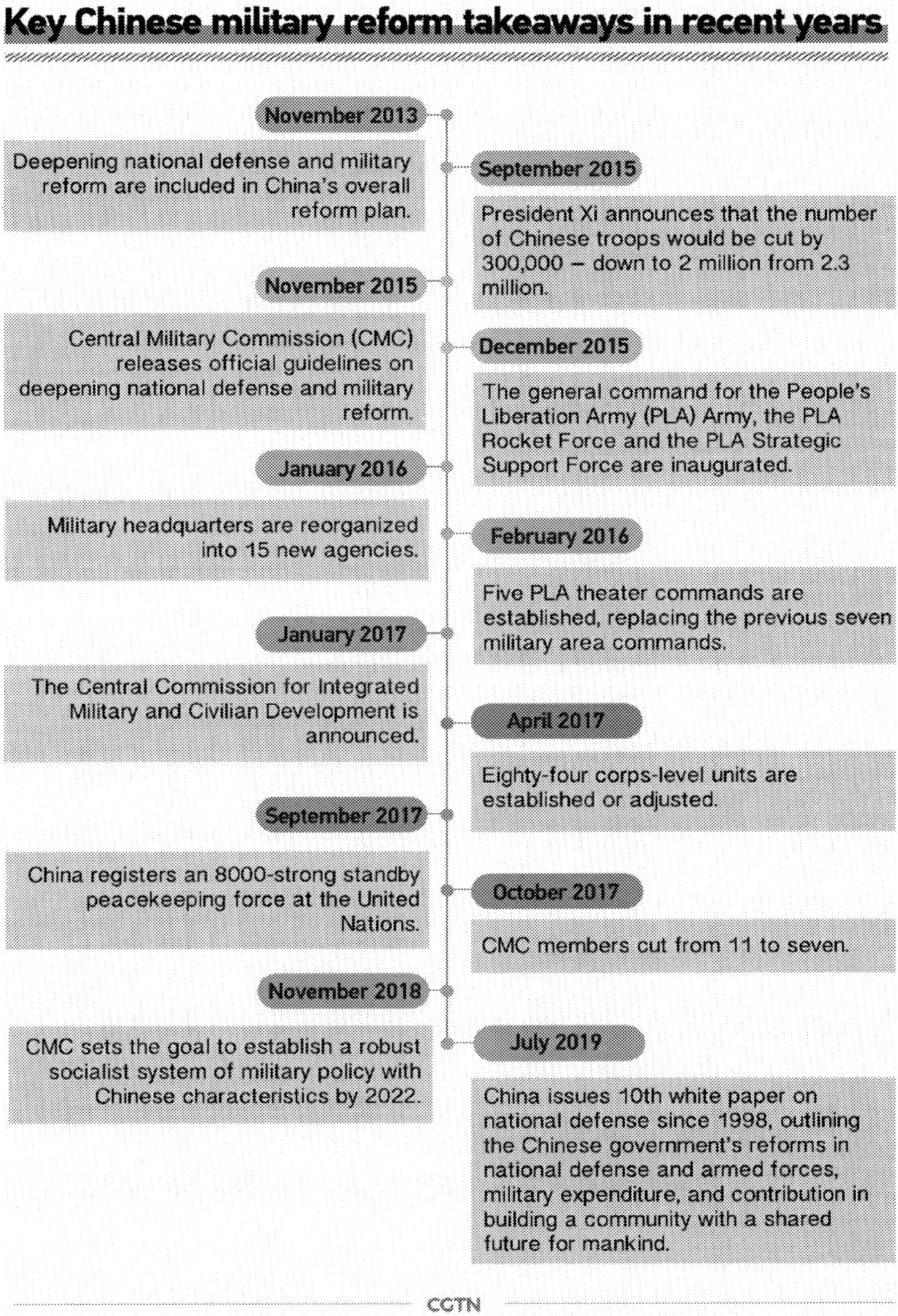

Image Source: ***China Global Television Network (CGTN)***

reorganization of military personnel, stressing quality and effectiveness over quantity of staff.

Reform in China's National Defence and Armed Forces

In September 2015, the Central Military Commission (CMC) released an official guideline on deepening national defence and military reform. Regarded as one of the biggest military reforms in modern China, the latest military reform has been on the agenda for China's rejuvenation in the new era since the 18th Communist Party of China (CPC) National Congress, in 2012. After the reforms, Xi Jinping is the general secretary of the Communist Party of China, President of PRC, and chairperson CMC, and he directly controls China's foreign relations, and the military. Some key developments in recent years are covered below.[2]

A series of measures were taken to remove institutional barriers and solve structural and policy-related problems. To enhance the operational effectiveness of the military, four general departments within the CMC

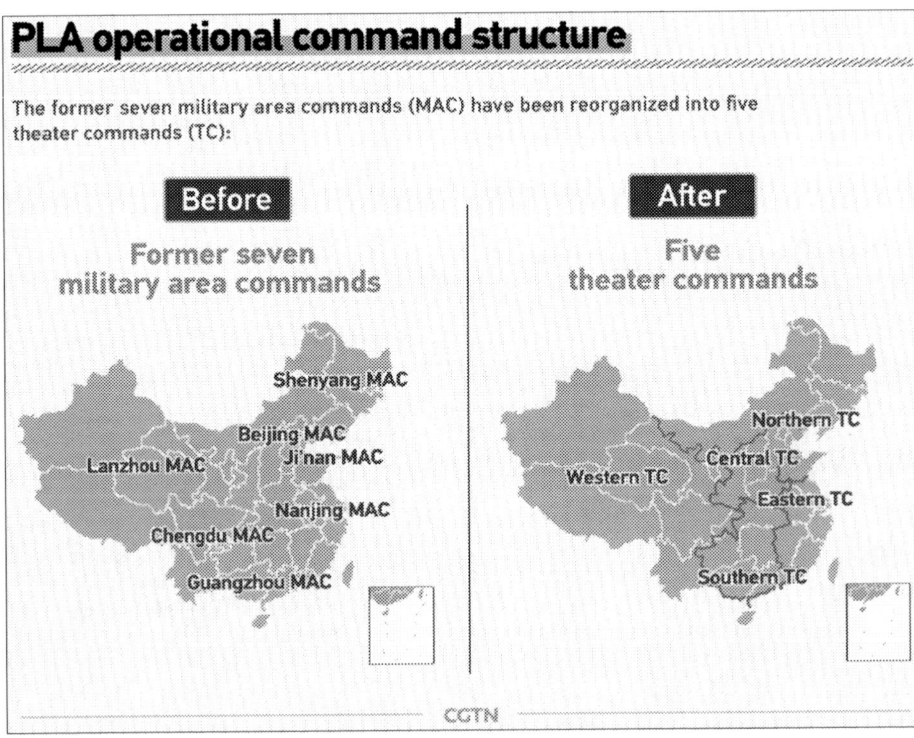

Image Source: ***China Global Television Network (CGTN)***

were reshuffled into 15 agencies under the CMC leadership to streamline the chains of command. The PLA Strategic Support Force was also created by combining strategic support forces across the services and CMC organs. The PLA Joint Logistic Support Force was established to enhance integrated joint operations capacity. The Second Artillery Force, which commands China's missile forces, was reorganised as the PLA Rocket Force. The seven military area commands were disbanded and have been reorganized into five theatre commands, named by their strategic locations in north, south, east, west, and central China. The new PLA operational Command Structure is considered operationally more viable.[3]

To optimize the size and structure of the army, adjust and improve the proportion between various troops, and reduce non-combat institutions and personnel, the Chinese military cut 300,000 personnel to keep the total active force at two million. The PLA has significantly downsized the active force of the PLA to less than 50 per cent of the armed forces. The active force of the PLAAF maintained its number, the PLAN and PLARF were moderately increased, and the force structures of all services and arms were optimized. As per Military Balance 2015 and 2019, the PLA numbers are shown below.

Services	*Number of Personnel*	
	2015	*2019*
Ground Forces	1,600,000	975,000
Navy	235,000	250,000
Air Force	398,000	395,000
Rocket Force	100,000	120,000
SSF	N/A	145,000
Other	N/A	150,000
Total Active Personnel	2,333,000	2,035,000
Paramilitary	660,000	660,000
Reserve	510,000	510,000

Information Source: Military Balance 2015 and 2019.

Defence White Paper 2019

China's new 2019 defence 'White Paper'[4] says international strategic competition is on the rise. The USA has adjusted its national security and defence strategies, and adopted unilateral policies. It has provoked and intensified competition among major countries, significantly increased its defence expenditure, pushed for additional capacity in nuclear, outer space, cyber and missile defence, and undermined global strategic stability. NATO has continued its enlargement, stepped up military deployment in Central and Eastern Europe, and conducted frequent military exercises. It says the region has become the focus of major country competition, bringing uncertainties to regional security. The USA is strengthening its Asia-Pacific military alliances and reinforcing military deployment and intervention, adding complexity to regional security. The deployment of the Terminal High Altitude Area Defence (THAAD) system in the Republic of Korea (ROK) by the USA has severely undermined the regional strategic balance and the strategic security interests of regional countries. It describes Taiwan, Tibetan, and Turkistan separatists as threats, the dangers of disputes over territory, and again highlights the U.S. by noting that, "Countries from outside the region conduct frequent close-in reconnaissance on China by air and sea, and illegally enter China's territorial waters and the waters and airspace near China's islands and reefs, undermining China's national security." The White Paper has a long section on "Reform in China's National Defence and Armed Forces." Military training in real combat conditions across the armed forces is in full swing. Since 2012, China's armed forces have carried out extensive mission-oriented training tailored to the specific needs of different strategic directions and exercises of all services and arms, including 80 joint exercises at and above brigade/division level. The Theatre Commands have strengthened their leading role in joint training and organized serial joint exercises to improve joint combat capabilities. The PLAAF has conducted combat patrols in the South China Sea and security patrols in the East China Sea, and operated in the West Pacific.

Air Assets at the 70th Anniversary Parade[5]

The PLAAF and the PLAN, for the first time in PLA history, formed a "joint squadron in the Support Aircraft Echelon" requiring improved coordination of the platforms of Y-9 and Y-8 in the air. The PLAN's Y-8 electronic reconnaissance aircraft was flown alongside the PLAAF's Y-8 jammer and Y-8 electronic counter-measures (ECM) aircraft. Although it appears to be a rather small step towards "joint operations" between the PLAAF and the PLAN, the symbolic meaning of flying PLAAF and PLAN special mission aircraft in a single formation should not be underestimated. The debut of the Y-9 communications counter-measures aircraft also demonstrated that the PLAAF is steadily improving its EW capabilities. The Y-9 is designed to "provide electromagnetic support for operations through long-range jamming, degrading enemy command, and control, and destroying enemy battlefield situation awareness", and it symbolizes that the development of the PLA's EW capability has "entered a new phase." The fly-past included three unmanned formations offering a glimpse of a variety of autonomous systems as well as possible command and control structures of such systems. Each formation comprised a large number of UAVs. The PLARF as a "Strategic Service" was on parade for the first time. There was an expansive display of the PLA's conventional and nuclear missiles. These included CJ-100 Cruise Missiles, DF-26 Nuclear and Conventional Missiles, DF-31AG Nuclear Missiles, DF-5B Nuclear Missiles, and DF-41 Nuclear Missiles.

Military Doctrinal Shift

The Maoist military doctrine which asked for a "combination of millet and rifles", changed after 1978, to Local War with its emphasis on active defence.[6] Jiang Zemin, brought in to "implement a strategy of active defence" in 1997, which came from lessons of the Gulf War, that a nuclear world war was unlikely.[7] The quest for offensive capabilities in addition to defensive was the natural consequence. In addition, the Chinese seem to outline the need for a quantitatively smaller but technologically superior force in the future.

PLAAF Doctrinal Shift

A great assessment of modern Chinese air power can be found in a 2011 monograph put together by a group of Rand researchers titled 'Shaking the Heavens and Splitting the Earth.'[8] The five basic concepts of executing strategic coercion, independent and concentrated use of air power, joint application with other services, strategic force delivery and seizing information and electromagnetic superiority reveals a new and aggressive intent. What it also indicates is that the Chinese were willing to project air power as a strategic tool only after spectacular advances in Chinese aviation and space technologies manifested into platforms and capabilities.

The PLAAF is the largest air force in the region and the third largest in the world, with more than 2,500 total aircraft (not including UAVs or trainers) and 1,700 combat aircraft. The PLAAF is closing the gap with Western air forces across a broad spectrum of capabilities such as aircraft performance, command and control, and electronic warfare. In 2017, the PLAAF reorganized its force structure as part of broader PLA reforms.

The changes included establishing at least six new airbases and restructuring the force's previously subordinate regiments into brigades under these newly established bases by disbanding fighter and fighter-bomber divisions. The PLAAF also relocated and subordinated some units to different theatre commands and re-designated the 15th Airborne Corps as the PLA Airborne Corps.

PLAAF Roles and Mission

In line with the strategic requirement of building air-space capabilities and conducting offensive and defensive operations, the PLAAF is endeavouring to shift its focus from territorial air defence to both defence and offense and build an airspace defence force structure that can meet the requirements of informatized operations. The PLAAF will boost its capabilities for strategic early warning, air strike, air and missile defence, information counter-measures, airborne operations, strategic projection, and comprehensive support.[9]

PLAAF BROAD COMPOSITION

As per the U.S. Defence Intelligence Agency Report 2019,[10] China has significantly modern air force assets.

Bombers

China's bomber force comprises variants of the H-6 Badger bomber, and the PLAAF has worked to maintain and enhance their operational effectiveness. The H-6K variant, which China fields in greater numbers, integrates standoff weapons and features more efficient turbofan engines in redesigned wing roots. This extended-range aircraft can carry six ALCMs, providing the PLA a long-range, standoff, precision-strike capability. The PLAAF is developing new medium-and long-range stealth bombers to strike regional and global targets. Stealth technology continues to play a key role in the development of these new bombers, which probably will reach initial operational capability no sooner than 2025. These new bombers will have additional capabilities, with full-spectrum upgrades compared with current operational bomber fleets, and will employ many fifth-generation fighter technologies in their design.

Early Warning

AEW&C aircraft, such as China's KJ-2000, KJ-200, and KJ-500, are force multipliers, amplifying the capabilities to detect, track, and target threats. These aircraft extend the range of a country's integrated air defence system network. In particular, these systems are better suited to detecting low-altitude targets at greater standoff distances. AEW&C aircraft also incorporate state of-the-art radar technology, such as AESA radars that offer instantaneous target updates, electronic beam steering, advanced radar modes, very large search volumes, and the ability to detect and track hundreds of targets simultaneously with increased ability to detect low-observable targets.

Fighters

Although the PLAAF still operates a large number of older third-generation fighters, it will probably become a majority fourth-generation force within the next several years. The PLAAF has fielded at least 600 fourth-generation fighters and is already inducting fifth-generation fighters.

Fourth-generation fighter aircraft include the Chinese J-10B/C, J-11B, and J-16, and generally have electronically or mechanically scanned multimode radars, passive infrared search and track systems, 'Glass' cockpits with multifunction displays (MFDs), improved heads-up display (HUD), helmet-mounted sight (HMS), high-bandwidth communications and data-links and identification, friend or foe (IFF), advanced EW avionics, including digital jamming system, radar warning receiver, chaff/flare dispensers, and adaptive countermeasures, engines with increased thrust and service life, advanced weapons, including long range AAMs, off-bore-sight short-range AAMs, ALCMs, and precision-guided munitions (PGMs). Some have passive electronically scanned array or AESA radars. These radars provide long-range radar detection and electronically scanned radar beams that enable automatic target acquisition, tracking of multiple targets, and highly accurate targeting data for air-to-air and precision air-to-ground engagements. Digital radio frequency memory (DRFM) jammers enabling instantaneous smart jamming responses by automatically selecting jamming waveforms to counter a specific radar threat—significantly improving fighter aircraft survivability.

The fifth-generation fighter aircraft are the Chinese J-20 and FC-31/J-31. These have stealth aircraft designs with significantly reduced radar and infrared signatures. AESA radars, long-range, multiband electro-optical (EO) targeting systems, fused sensors, advanced glass cockpits with large MFDs and HMSs, advanced data-links fusing data from air and ground networks, internal carriage of off-bore-sight and long-range AAMs, ALCMs, and PGMs, sophisticated EW suites with advanced DRFM jammers and EO defensive systems. They also have super manoeuvrability and/or super cruise capability. They are designed with network-centric warfare technology, and potent air-to-air lethality and standoff attack capabilities.

UAVs, UCAVs and Drones

The huge Chinese UAV industry includes a large number of small start-up companies with newer innovative models. These include stealth UAVs, morphing UAVs, annular wing VTOL design micro-UAVs, unmanned airships, flying wings, modified light aircraft and sailplanes, UFO-style flying discs and even ornithopters (utilising flapping wings). The ASN Technology Group is the biggest UAV producer in China. Over 40 different types of UAVs have been manufactured and over 1,500 delivered by ASN.

China's has a growing high-end military drone force. Recent years have seen the introduction of a variety of new UAVs in service as well, some of which are equivalent to Western ones. Medium Altitude Long Endurance drones have a service ceiling of below 9,000 metres and over 24 hours endurance or longer. The PLAAF's current in-service MALE drone fleet consists of the BZK-005 (Giant Eagle) with an electro-optic (EO) sensor turret. The GJ-1 and GJ-2 are MALE UAVs capable of a strike role and are variants of the original export oriented Wing Loong I and Wing Loong II systems. Both GJ-1 and GJ-2 can launch the KD-9/10 laser guided anti-tank guided missiles. The larger GJ-2 can carry a larger payload and has a chin mounted synthetic aperture radar with longer target acquisition range. GJ-1 and GJ-2 are considered the equivalents of MQ-1 and MQ-9, respectively.

High Altitude Long Endurance (HALE) drones with a service ceiling close to 18,000 metres, are powered by jet engines. The RQ/MQ-4 Global Hawk and Triton family of the USA are in this category. PLAAF HALE is WZ-7, a box wing design, with twin tails and a dorsally mounted jet engine. WZ-7s are deployed in various airbases around China, and were deployed in Tibet during the Doklam standoff with India in 2017. The other PLAAF HALE drone is the large, twin joined hull aircraft commonly called Divine Eagle. The aircraft is thought to be an Airborne Early Warning (AEW)

UAV. The joined hull allows installation of conformal radar arrays on the sides and perhaps the front surfaces of the fuselage. The Divine Eagle is reportedly still under development.

The GJ-11 stealth UCAV and the WZ-8 supersonic reconnaissance UAV were both unveiled at the 2019 National Day parade. The GJ-11 (Sharp Sword), the first prototype flew in November 2013. The WZ-8 is a supersonic reconnaissance UAV with some sort of rocket-based propulsion system. The PLAAF is seeking to close the technological and capability gap with the USA. The GJ-11 is a stealth UCAV with a deep strike purpose, and estimates of its dimensions would make it slightly larger than the European Neuron demonstrator and about the size of the Boeing Phantom Ray demonstrator. Such a UCAV would likely have a combat radius of over 1,000 km, and its internal weapons carriage bays should be able to accommodate either two 500 kg guided bombs or eight 100 kg guided bombs. The GJ-11 could be used to develop concepts such as autonomous missions, autonomous swarming, manned-unmanned-teaming (MUMT), loyal wingman, and other concepts. The Divine Eagle AEW UAV offers to significantly augment the existing manned AEW&C platforms, allowing more persistent AEW coverage. AEW UAVs could also be deployed in battle spaces too dangerous for a manned AEW&C. The "Anjian" (Dark Sword) UCAV is a delta winged, twin-tail, and canard equipped platform. Details are not clear, but some reports suggest that it could be a Global Hawk class HALE UAV. The WZ-2000 is a twin-jet powered delta winged high-altitude long-endurance UAV similar in size to the U.S. General Atomics MQ-9 Reaper. China plans to outpace U.S. spending on unmanned systems and plans eleven drone bases on the coastline for marine surveillance in disputed areas of the South China Sea. In due course, China will induct many UAVs/UCAVs. The Wing Loong, is a Medium-Altitude Long Endurance (MALE) UAV intended for use as a surveillance and aerial reconnaissance platform. It is capable of being fitted with air-to-surface weapons for use in an unmanned combat aerial vehicle (UCAV) role. It can carry the BA-7 air-to-ground missile, YZ-212 laser-guided bomb, YZ-102A anti-personnel bomb and 50-kilogram LS-6 miniature guided bombs. Wing Loong II is an upgraded variant with provisions for up to twelve air-to-surface missiles, and officially entered service with the PLAAF in November 2018.

In September 2011, the Aviation Industry Corporation of China (AVIC) had sponsored the first nationwide international UAV innovation Grand

Prix. In 2015, another event was held, which generated many ideas. They included rotary UAVs, supersonic, super maneuverable and low observable platforms for air-to-air missions, models of forward swept wing, twin tailed, single engine, canard equipped aircraft, jet engine UAVs, etc. The Wing Loong is a well-established armed UAV with laser-guided HJ-10, a hellfire class of anti-tank missile. The CH-3 is a fixed wing UCAV with canard layout. The CH3 UAV carries an Anti-Radiation Missile (ARM), and an Air-to-Surface Missile (ASM) with a Semi-Active Laser (SAL) seeker, suitable for Close Air Support (CAS) missions. Some inertial/GPS (INS/GPS) guided munitions have also been produced for CH3. The Pakistani Burraq UAV is based on the CH3 UAV and has been supported by China. China is also developing a wide spectrum of electro-optical payloads with attachable turrets. These include medium range Forward Looking Infra Red (FLIR), IR array systems, radar payloads, small Ku band Synthetic Aperture Radars (SARs), real time image processing techniques, etc. They are also interfacing existing UAV earth stations with satellite communication stations for seamless connectivity and data-sharing.

In recent times, the PLAAF has improved its UAV capabilities, unveiled new aircraft that combine strike capability with aerial reconnaissance, and deployed select platforms to new areas, such as the South China Sea. China's first turbine-powered armed reconnaissance UAV is the Yunying (Cloud Shadow), and the armed ISR UAV Gongji. Marketing material for China's armed UAVs cite speeds of 275 kmph, endurance of 20 hours, and payloads of two or more air-to-surface guided munitions.

Aerial Weapons

Besides conventional payloads, the PLAAF is developing smart, intelligent, and precision ammunition with surgical strike capability. The KD-88 is an IR/TV guided Air-to-Surface Missile (ASM) for ground and sea targets with an estimated range of about 180 km. The other state-of-the-art arsenal in use are the supersonic Russian and Chinese-made Anti Radiation Missiles (KH 31P, YJ-91) for operational use with SU-30, JH 7A and J-11. There is an inventory of laser-guided and satellite-guided bombs. All 4th generation PLAAF aircraft can carry PGMs. The top end PLAAF Beyond Visual Range (BVR) Air-to-Air Missiles (AAM) includes the latest PL-12 and PL-21 (Comparable to the U.S. AIM-120 AMRAAM, and the Russian R-77). Pakistan has ordered 600 PL-12s. The latest PL-15, very long range AAM, which entered service in 2016, is equipped with an active electronically scanned array (AESA) radar. It is carried by J-20, J-11B, J-16, and J-10C fighters and likely to be exported to Pakistan for JF-17 Thunder.

Currently China has up to 500 DH-10 land attack cruise missiles with a 1,500-km range, and part of these is air launched. It has the R-27 semi-active radar/infrared, short-range infrared guided R-73, and the active radar homing R-77, all of which are part of the IAF inventory. China has also been upgrading its air combat capability with a variety of home-grown air-to-air missiles, including the extended-range radar-guided PL-15 that can threaten force multipliers like air-to-air refuelling tankers and Airborne Early Warning and Control Systems (AEW&C).

Surface-to-Air Missiles

The PLAAF has one of the largest forces of advanced long range SAM systems in the world, made up of a combination of Russian-sourced SA-20 (S-300PMU1/2) battalions and domestically produced CSA-9 (HQ-9) battalions. In early 2018, Russia began delivering to China the S-400/Triumf SAM system, which Beijing will use to improve its strategic air defence systems as a follow-on to the SA-20. The PLAAF may simultaneously develop its domestically produced CSA-X-19 (HQ-19) to provide the basis for a ballistic missile defence capability. The PLAAF purchased the Russian S-300 surface-to-air missiles (SA-10 and SA-20) and produced the comparable indigenous HQ-9. PLAAF has about 192 modern launchers along with 490 legacy launchers. China has acquired six batteries of the game-changer S-400 air defence systems. Delivery of the system began in January 2018. China test-fired the Russian S-400 system for the first time in early August 2018.

Transport

China's aviation industry continues to test its Y-20 large transport aircraft for introduction into the PLA's operational inventory to supplement and eventually replace China's small fleet of strategic airlift assets, which includes a limited number of Russian Il-76 aircraft. The Y-20 conducted its maiden flight in early 2013 and reportedly uses the same Russian engines as the Il-76. The large transports are intended to support airborne command and control, logistics, para-drop, aerial refuelling, and strategic reconnaissance operations as well as Humanitarian Assistance and Disaster Relief (HADR).

C4ISR

The PLAAF is conscious of the need for Command, Control, Communications, Computers, Intelligence, Surveillance, and Reconnaissance (C4ISR) systems and China is developing or acquiring systems and platforms. By 2020, the PLAAF probably hopes to develop an ISR capability to effectively support traditional air missions, including ground support and air superiority, along with the PLA's emerging military capabilities in space. This essentially means seamlessly integrating all of its air and space capabilities, which is a significant challenge.

Airborne Corps

As part of the ongoing military reforms, in 2017 the PLAAF reorganized its airborne units into a new Corps, the PLA Airborne Corps, consisting of six airborne brigades, a special operations brigade, an aviation brigade, and a support brigade. Previously, airborne forces fell under the PLAAF 15th Airborne Corps and were organized in three airborne divisions, supported by a special operations group, with both fixed and rotary-wing assets. The PLA Airborne Corps' main purpose is to carry out both airborne and air-assault operations. These operations are designed to support main-force operational efforts, seize, and hold key targets and areas in the enemy's depth such as airfields and bridges, block an enemy's retreat, block reinforcement by enemy reserve forces, and conduct raids on key targets in the enemy's depth.

Brigadisation of PLAAF[11]

The reforms initiated by Xi Jinping in 2016 continue to have significant effects on the PLAAF in its missions, organizations, and its people. One of the most evident of these changes is the increased shift toward a brigade system. The PLAAF has rapidly pursued a program of shifting from the traditional Division-Regiment organization, to a new Base-Brigade structure for its fighter and ground attack aircraft. It is being implemented across the PLAAF, with the current exception of the bomber and special mission aircraft organizations. Since 2015, the PLA Air Force and PLA Navy Aviation long-distance over-water training has become more frequent, featuring a greater variety of aircraft, and expanded in geographic scope.

In late 2011, the PLAAF first began to restructure its seven flight colleges into three flight colleges and one Flight Instructor Training Base. It also began to restructure its operational fighter and ground attack air divisions and subordinate regiments into a brigade structure. In January 2012, the PLAAF began creating four Corps deputy leader-grade bases from the existing command posts. Each base is responsible for command and control of the air brigades, SAM, AAA, and radar units in their AOR. They also coordinate with Army and Navy units in their AOR for joint training. Each of the newly-created 14 operational air brigades was re-subordinated under four newly reconfigured Corps deputy leader-grade bases. These bases were previously corps deputy leader-grade or division leader-grade command posts that were, in turn, created in 2004. However, after the initial restructuring in 2011-2012, no further changes to any fighter and ground attack units or operational bases were made until 2017. In early 2016, as part of the "above the neck reforms" (a term the PLA uses to refer to units above the corps level), the PLAAF Headquarters completely reorganized its administrative and functional departments, and merged its seven Military Region Air Force (MRAF) Headquarters into five Theatre Command Air Force (TCAF) Headquarters to align with the overall PLA transition from a Military Region structure to a Theatre Command structure. The 19th Party Congress, in late 2017, reorganized the PLA's Central Military Commission (CMC). Post reorganization, the service commanders were no longer included in the CMC.

It is important to note that the use of the word "base" here is in the Chinese context of an organizational unit at the Corps deputy leader grade, not in the American sense of a physical location. In 2017, as part of the "below the neck reforms" (i.e., Corps and below), the PLAAF completed its reorganization of all remaining combat fighter and ground attack divisions and regiments into brigades and placed them under several newly configured bases from previous command posts. The PLAAF has not yet reorganized its existing three bomber divisions, three transport divisions, or its three specialized air divisions into a brigade structure. Given that these units are either directly subordinate to PLAAF HQ or to a TCAF HQ, and not a base, it is not clear if this restructuring will happen in the future. The PLAAF has created six Transport and Search and Rescue (SAR) Air Brigades. The PLAAF has also organized other types of organizations into a brigade structure. These include unmanned aerial vehicle (UAV) brigades, Flight Test and Training Base Air Brigades, and Transition Training Base Air Regiments and Brigades. The few bases that retain training regiments are, however, likely to keep them. The PLAAF Airborne Corps, which was formerly known as the "15th Airborne Corps" has transitioned from having subordinate divisions and regiments to having only subordinate brigades. These include a special operations brigade and support brigade. It also has a newly-created Aviation (transport) Brigade. It is not clear if the Chinese Aerospace Studies Institute remains as a regiment, is upgraded to a brigade, or incorporated into the aviation brigade as multiple flight groups. The PLAAF has created several unmanned aerial vehicle (UAV) brigades and is likely to add more units in the near future. The current reorganization of the PLAAF is ongoing, and many of the details are still developing and are simply unknown as yet.

Although the services are now more equal in terms of organizational structure and hierarchy, the new commanders and political commissars of the five theatre commands still come from the ground forces. This implies continued Army dominance within the system. After the current "transition" leadership of the new theatre commands eventually retire, one area to watch will be how Navy and Air Force officers are integrated into the leadership of the commands.[12]

PLAAF Operational Equipment Holdings

As per Military Balance 2020,[13] the PLAAF has 395,000 personnel. It has at least six new air bases, and is restructuring previously subordinate regiments into brigades under the newly established bases by disbanding its fighter and fighter-bomber divisions. A typical PLAAF Brigade (Bde), in some fleets still called Regiment (Regt) is known to be assigned between 25 to 32 aircraft, with one or two brigades per airfield. Its current known holdings of bomber aircraft include one Regt of H-6M, two Regt of H-6H, five Regt with H-6K, and one Bde with H-6N still being formed. Thus there are total of nine Bomber brigades. The Fighter fleet includes five Bdes each of J-7 and J-7E, three Bdes of J-7G, four Bdes of J-8F/H, two Bdes of J-11A/Su-27SK/Su-27UBK, four Bdes of J-11A/J-11B/Su-27UBK, and three Bdes of J-11B/BS, totalling 26 brigades. The Fighter/Ground Attack fleet comprises 21 brigades, of which seven Bdes of J-10A/S, one Bde of J-10A/C/S, two Bdes of J-10B/S, one Bde of J-10B/C/S, two Bdes of J-10C/S, one Bde of Su-35, three Bdes of J-16, three Bdes of Su-30MKK, and one Bde of J-20A is still under formation. There are six Bdes of JH-7A dedicated ground attack aircraft. The total fighter/bomber brigades are 62.

The two Regts of Y-8CB/G/XZ are for Electronic Warfare. One Regt each JZ-8F and Y-8H1, and one Bde with JZ-8F are used for ISR. There are two Regts of AWACS and AEW&C aircraft comprising KJ-200, KJ-500, KJ-2000, and Y-8T. For Search and Rescue, there are four Bdes and one Regt with Y-5, Mi-171E, and Z-8. One Bde of FRA is with H-6U.

There are two VIP Regts with B-737, CRJ-200/700, Tu-154M, and Tu-154M/D aircraft. One Regt with AS332 Super Puma and H225 helicopters is also for VIP duties. The rest of the Transport aircraft fleet consists of one Regt of Il-76MD/TD, and one Regt with Il-76MD and Il-78. There are eight other Regts for transportation including one Regt of Mi-17V-5, one Regt with Y-5/Y-7/Z-9, one Regt with Y-5/Y-7, three Regts with Y-7, one Regt with Y-8, and one Regt with Y-8/Y-9.

For training, the PLAAF has five Bdes of CJ-6/6A/6B, Y-5, eight Bdes of J-7, JJ-7A, and 13 Bdes with JL-8, JL-9, and JL-10. For ISR they have two Bdes with GJ-1 and GJ-2 UAVs, For air defence, the PLAAF has 23 SAM Bdes.

The PLAAF has 2,517 combat capable aircraft. Type-wise holdings are:

1. **Bombers** (176) - 12 H-6A (trg role); 60 H-6H/M; 100 H-6K; around 4 H-6N.
2. **Fighters** (759) - 200 J-7; 192 J-7E; 120 J-7G; 50 J-8F; 50 J-8H; 95 J-11; 20 Su-27SK; 32 Su-27UBK.
3. **Fighter/Ground Attack (FGA)** (794+) - 220 J-10A; 55+ J-10B; 100+ J-10C; 70 J-10S; 130 J-11B/BS; 100+ J-16; 22+ J-20A (entering service); 73 Su-30MKK; 24 Su-35.
4. **Attack** - 140 JH-7A.
5. **EW** (14) - 4 Y-8CB; 6 Y-8G; 2 Y-8XZ; 2 Y-9XZ.
6. **ELINT** 4 Tu-154M/D.
7. **ISR** (51) - 24 JZ-8; 24 JZ-8F; 3 Y-8H1.
8. **AEW&C** (13) - 4 KJ-200; 5 KJ-500; 4 KJ-2000.
9. **C2** (5) - 2 B-737; 3 Y-8T.
10. **FRA** (13): 10 H-6U; 3 Il-78.
11. **TPT** (336) - **Heavy** (28+) - 20 Il-76MD/TD; 8+ Y-20; **Medium** (42+) - 30 Y-8C; 12+ Y-9; **Light** (239) - 170 Y-5; 41 Y-7/Y-7H; 20 Y-11; 8 Y-12; **Passenger** (27) - 9 B-737 (VIP); 5 CRJ-200; 5 CRJ-700; 8 Tu-154M
12. **Training** (1,012+): 400 CJ-6/-6A/-6B; 12+ HY-7; 50 JJ-7; 150 JJ-7A; 350 JL-8; 30 JL-9; 20+ JL-10.

13. **Helicopters - Multirole (**22): 20 Z-9; 2 Mi-17V-5; **Transport (**31+) - **Heavy** 18+ Z-8; **Medium (**13+) - 6+ AS33 (VIP); 3 H225 (VIP); 4+ Mi-171.
14. **UAV - CISR - Heavy (**12+) GJ-1; GJ-2; GJ-11 (under testing) - **ISR - Heavy (**14+) - 12+ EA-03; 2+ WZ-8.

PLAAF Air Defence assets are:

1. **SAM (**850+), **Long-range** 516+ 180 HQ-9 (CH-SA-9); 60 HQ-9B (CHSA- 21); 100+ HQ-22; 32 S-300PMU (SA-10); 64 S-300PMU1 (SA-20); 64 S-300PMU2 (SA-20); 16 S-400 (SA-21B). **Medium-range (**230): Around 80 HQ-2/-2A/-2B (CH-SA-1); 150 HQ-12 (CH-SA-12) **Short-range (**104+): 50+ HQ-6A (CH-SA-6); 24 HQ-6D (CH-SA-6); around 30 HQ-7 (CH-SA- 4) .
2. **GUNS-**16,000 100mm/85mm.
3. **AAM -** PL-5B/C; PL-8; R-73 (AA-11A); PL-10; R-27 (AA-10); PL-11; PL-12; PL-15 (entering service); R-77 (AA-12A); R-77-1 (RVV-SD) (AA-12B)

Air-to-Surface Weapons are:

1. AKD-9; AKD-10; KD-88; Kh-29 (AS-14); Kh-59M (AS-18); Kh-31A (AS-17B); Kh-31P (AS-17A); YJ-91 (Domestically produced Kh-31P variant); ALCMs CJ-20 and YJ(KD)-63; Laser Guided Bombs LS-500J; LT-2; TV-Guide KAB-500KR and KAB-1500KR.

Airborne Corps - One special operations bde; six Airborne bdes; one heliborne regt with eight WZ-10K, eight Z-8KA, and 12 Z-9WZ; one support bde; one transport bde with two Y-7, six Y-8. 12 Y-12D, Point Defence SAMs (CH-SA-7).

Air Assets Western Theatre[14]

In the Western Theatre Air Force opposite the Indian border across the Himalayas, PLAAF assets include the **4th Transport Division** having one Transport regt with Y-8/Y-9; one Transport regt with Y-7; and one Transport regt with Mi-17V-5/Y-7/Y-20. **Lanzhou Base** has one fighter bde with J-11AB; one fighter bde with J-7; one fighter bde with J-7E; 1 FGA bde with J-16; 1 SAM bde. **Urumqi Base** has one fighter bde with J-8H; one fighter bde with J-11A/B; one attack aircraft bde with JH-7A; 2 SAM bde. **Lhasa Base** has one SAM bde. There are sizeable air asset, at Hotan airbase which is at an altitude of 4,500 feet.

PLAAF Major Aircraft Assets (As per Military Balance 2020)

Type	*Origin*	*Class*	*Role*	*Numbers*
J-7 Variants	China	Jet	Fighter/Trainer	852+
J-10	China	Jet	Multi-role Fighter	465+
J-11A, B, BS	China	Jet	Air Superiority Fighter	225
J-8	China	Jet	Fighter	100
JL-8	China	Jet	Trainer	350
Su-27	Russia	Jet	Air Superiority Fighter	52
Su-30 MKK	Russia	Jet	Multi-role	73
Su-35S	Russia	Jet	Multi-role	24
H-6 Variants	China	Jet	Bomber	176
IL-76	Russia	Jet	Transport	20
Y-7	China	Propeller	Transport	41
Y-8	China	Propeller	Transport	30
Y-9	China	Propeller	Transport	12+
Y-20	China	Jet	Transport	8+
IL-78 MD/TD	Russia	Jet	Aerial tanker	3
Tu-154M	USSR	Jet	Patrol/ELINT	8
KJ-200	China	Propeller	AEW&C	4
KJ-500	China	Propeller	AEW&C	5
KJ-2000	China	Jet	AEW&C	4
Z-8	China	Rotorcraft	Transport	18+
Z-9	China	Rotorcraft	Transport	20
Z-18	China	Rotorcraft	Transport	Inducting
Z-20	China	Rotorcraft	Transport	Inducting
Mi-17	Russia	Rotorcraft	Transport	6

Note: PLAAF is known to have inducted unspecified numbers of J-20 recently.

PLAAF Hierarchical Stature

A 2017 edited volume from the China-centric Jamestown Foundation in Washington DC titled 'China's Evolving Military Strategy' is an excellent primer to look at the current PLAAF doctrine and strategy based on multiple Chinese strategic prescriptions from the Academy of Military Sciences and the National Defence University.[15] At the fundamental level, there is much that has been written in the 'Science of Military Strategy' (2013) that challenges many of the "sacred cows of the PLA, starting with the dominance of the ground forces and calling for a more equal focus on ground, sea, air, space and cyber domains." Space has now emerged as a domain that falls under the new Strategic Support Force, albeit still under the knowledge-driven umbrella of the PLAAF. This doctrinal shift much resembles the IAF's dilemma when it comes to claiming ownership of Space, but having to share control with the other two services as much bandwidth is used by all the three services. Just as a tri-service Defence

Space Agency has evolved in India, the new Strategic Support Force is almost ready to assume ownership of Space in China with an initial 'PLAAF heavy' structure, but with shared ownership.

The second and most striking development that ought to be taken seriously is the presence of two high-ranking PLAAF officers on the Central Military Commission (CMC).[16] While the PLAAF chief was granted a slot on the CMC along with the PLAN and PLARF chiefs in 2004, it is the elevation of a former PLAAF commander Xu Qiliang in 2012 as a Vice Chairman of the CMC that has indicated Chinese President Xi Jinping's resolve to develop air and space power capability at a rate that could take the world by surprise in a decade or so. It is also an indication of Jinping's willingness to look beyond the PLA for military guidance.

Key PLAAF Strategic Takeaways

The emerging PLAAF capabilities, like modern air forces, is the emphasis on long range offensive precision strikes in air and space with enablers like FRA and AWACS that seem to focus on countering any aggression with a combination of air and surface launched weapons. This obviously calls for synergy between the PLAAF and the PLSRF.[17] The PLAAF's multi-layered air defence systems has the ability to promulgate and enforce Air Defence Identification Zones (ADIZs) and 'no fly zones'. This is visible in its eastern seaboard to cover well beyond the first island chain in response to the U.S. pivot to Asia and the emergence of the Indo-Pacific strategic maritime space. This would also imply the PLAAF's increased capability to cover vast airspace in TAR and look deep into Indian air space. China's extensive constellation of surveillance satellites with short revisit cycles adds significant punch to target locating and tracking capability. This is an area where China has a lead over India's indigenous military space program. The formation of India's Defence Space Agency and more military-dedicated satellites should reduce the gap.

The PLAAF's tactical and strategic airlift capabilities are still work in progress. Notwithstanding the excellent infrastructure in TAR, which supports the surface movement of formations and logistics support, the PLAAF is conscious of the IAF's increased responsiveness by way of its airlift capabilities and operationalisation of numerous Advance Landing Grounds (ALGs) and airstrips along the Line of Actual Control (LAC).[18] Fast induction of China's Y-20 will add to the PLA's ability to induct and switch forces. Finally, the PLAAF's increased focus on electronic warfare, cyber and other support forces like 'base protection forces' is testimony to a concerted drive to make the PLAAF a contemporary and modern force. The newly created PLA Strategic Support Force (SSF) also offers significant force multiplier support to the PLAAF, something the Indian armed forces are just beginning to act by harnessing space, cyber, and information warfare resources.

Army Centric to Independent Approach

After PLAAF started becoming independent of Army-centric thinking in the 1990s, campaigns were divided into two categories—defensive campaigns and attack campaigns—either of which could be independent air force campaigns, or a part of a joint campaign. By 1997, the Chinese air force had evolved service-specific strategies.[19] The PLAAF primary missions now on were air deterrence, air offensive, air blockades, and support for ground force operations. Air offensives would entail large-scale strikes with the goal of rapidly gaining air superiority, reducing an adversary's capacity for military operations, and creating conditions for victory. It would entail attacks on airfields and seaports as well as air, land, and sea transportation routes with the goal of cutting off an enemy. Support for ground force operations would include attacks on logistics facilities, and key choke points, such as bridges. It would also include battlefield close air support, strategic and theatre airlift, and airborne operations.

PLAAF operational principles included surprise and first strikes, concentrated application of best assets, offensive action as a component of air defence, and close inter-service coordination. Since 2004, the PLAAF shifted to "Integrated Air and Space" strategy.[20] It now began being seen as a truly independent service. It is now possible for the Air Force to conduct an air sub-campaign where the ground forces, Navy electronic warfare and special operations elements, and long-range surface-surface missile forces play a supportive role.

The PLAAF is looking at continuous offensive to deny enemy breathing space through a "quick battle and quick resolution" campaign to bring early finality.[21] It will use high speed, precision, and lethality of high-tech weapons to quicken the war process. However, the high cost of high-tech weapons also means that war can be too costly if it is allowed to drag on or escalate. Reliance on measures that may help to gain an "element of surprise" in campaigns include disseminating disinformation, camouflage and concealment, electronic interference and deception, applying anti-infrared, anti-radar, and other stealth technologies. Developing and deploying anti-surveillance satellite measures; damaging enemy ground and air surveillance radars, and unmanned drones; monitoring radios and other major sensors; and strikes by long-range missiles, attack aircraft, and special forces; and employing laser, kinetic energy, and particle beam weapons during early operations.

Chinese Air Campaign Strategy

The PLAAF's first core mission is to defend China's airspace, The Beijing Capital Region, the HQs of the five "Strategic Zones", and the Shenyang Military District in NE China bordering Russia. The second core mission is to prepare for an assault on Taiwan. The third is to acquire the capability to project power into the South China Sea. A typical air campaign will start with an information offensive, which would entail cyber network attacks, electronic deception, electronic interference, and firepower destruction. In parallel, penetration of enemy air defences would be initiated.

The Chinese air strategy is moving towards power projection through strategic airlift, offensive capability beyond Chinese borders and strong tactical presence in terms of modern fighter aircraft. A significant part of the Chinese defence budget is being spent on air and air defence assets. These include multiple aircraft programs, aircraft carriers, UAVs, S-400 class SAMs, and long range AAMs among other assets. Missiles are a significant factor of the military, and are currently undergoing an unflagging modernization process. Ballistic Missile technology travels in tandem with space technology.[21] Spectacular advances have been made by China in both these synergistically related fields. The fact that China's missile programs have benefited immensely from reverse engineering and acquiring designs through surreptitious means does not detract from the fact that China now possesses a formidable array of missiles encompassing the entire classification types. There is a profound relationship that Chinese missiles have with the air strategy that seems to guide the PLAAF.

The PLAAF has a 'Rapid Reaction Strategy', which includes pre-emptive first strike with a strong punch and offensive air operations.[22] Large Area Defence (LAD) will help defend cities, industry, and airbases. The forward edge of the battle is planned to be pushed towards the enemy. Gradually, the fixed defences are giving way to mobile air defences with shoot-and-scoot ability. Mobility also plugs holes in air defences. Offensive air defence relies on integrated attacks on enemy targets, while keeping a strong air defence punch ready to take on the opponent's counter attacks. The aim is to maintain an offensive posture, forcing the enemy into a reactive mode and ultimately seizing the operational initiative. Network-centric operations are meant to support such an offensive approach.

The PLAAF campaign strategy and tactics switched from limited war to offensive action using high tech weaponry. Its state-of-the-art systems include AEW, EW aircraft, and SAMs. The new weapon technologies for target acquisition, precision, lethality, stealth and the force projection potential allow "limited wars". For some time now, Chinese air strategy is to counter external threats from the air through a defensive pre-emptive strike capability against targets outside its own space. These are planned to be integrated deep strikes and concentration of superior firepower to destroy opponent's retaliatory capabilities. This pro-active doctrine essentially seeks to take the battle into enemy territory.

The PLAAF has switched from territorial air defence to defensive offensive capability. They have built a network-centric air space defence force structure, a strategic airlift capability represented by airborne divisions, and the formation of a Rapid Reaction Force as part of the offensive active defence. The PLAAF doctrine also points towards the unification of air and space defences requiring integrated command and control based on the understanding that whoever controls space, controls the planet. Joint air defence operations will require the building of synergy with surface and sea-based assets, by integrating the sensor, shooter, and battle management capabilities.

Training to Fight and Fight to Win Approach

Key findings of a 2016 Rand Corporation report[23] on PLAAF training titled "From Theory to Practice" gives an insight into flying training. The PLAAF has embarked on a major institutional change so as to train for "actual combat conditions", and to prepare to "fight and win" wars against potentially superior military competitors. The training as at the basic, intermediate and advanced levels to prepare pilots prior to their arrival at their operational units. Then there is the training for transition to the PLAAF operational aviation units. There are separate flying schools to train instructors and units to evaluate the operational competency and weaknesses within the PLAAF aviation training system. In addition, they have fighter and attack pilot training units. The PLAAF has professionalized regiment level training allowing freedom to exploit weapon platforms and to train for realistic battlefield scenarios. It seeks to cultivate more autonomous decision-making among fighter pilots and has recently begun to shift training away from an emphasis on ground control to a system that encourages decision- making by formation leaders. However, structural weaknesses persist, and aviators continue to struggle when coordinating individual unit autonomous operational training with other units. The PLAAF also lacks a unit analogous to the USAF Air Weapons School (AWS) or any other similar means of developing the combat skills of midcareer pilots. The top-down nature of military training guidance in the Chinese system produces common concepts throughout the PLA's services and branches but does not produce a high degree of "jointness." PLAAF units have an Annual Training Cycle. Normally a theme for the year is set at the Headquarter level. Theatre Air Force (TAF) brigades can also set operational training themes like "defence penetration, assault, and long-range precision strike combat methods". Aviation unit commanders adjust and tailor certain types of training activities in accordance with local issues and challenges.

Unrealistic training in the past manifested itself in multiple ways that hindered the PLAAF's air combat capabilities. The PLAAF recognizes that a gap exists between the skills of its pilots and those in the air forces of powerful nations. To address training weaknesses, it has begun 'training for battle' instead of just going through the motions. It conducts training and exercises intended to gauge its progression in conducting new missions and support cross-theatre operations associated with the PLA's transformation. The training objectives focus on improving the air force's offensive and defensive air capabilities, enhancing realistic combat training, improving joint training, and training over distant seas to move the force closer to achieving the goal of building a world-class air force.

During the past few years, the PLAAF has revamped its training program and now seeks to replicate real-world combat environments as closely as possible. PLAAF pilots spend most of the summer months participating in multi-disciplinary realistic operational exercises, drills, and evaluations, including inter-service exercises. "Red" and "Blue" teams operate under "complex electromagnetic environments", and the latest fighter aircraft are also used. Tactics employed reflect the recent shift from ground-controlled engagements to control by flight leads. Joint debriefs involve video playback, flight-data transmission feedback, and other electronic means. Exercises are meant to gauge "actual combat ability", not simply obtain "high marks." Estimates are that at least 60 per cent of all training at an operational unit is devoted to tactics training.

In the late Eighties, the PLAAF had formed the "Blue Army"—an aggressor unit composed of top-notch aircrew and high-end equipment and aircraft.[24] The 8th PLAAF Flight Academy is home to the Su-30MKK fighter-bomber "Blue-Army Aggressor Squadron." It is trained to act as the enemy in war-games by simulating the USAF F-15E or other Su-30 equipped Asian air forces. There are a total of three Blue-Army Aggressor squadrons within the PLAAF, the other two are equipped with J-10As and J-7Es to simulate lighter, single engine, opponents. These squadrons operate under the PLAAF Flight Test & Training Base (FT&TB) at Cangzhou-Cangxian Air Base in Hebei. During 1996, the Blue Army engaged PLAAF units in a two-day "offensive-defensive" exercise over the Gobi Desert.

Through its exercises named 'Red Sword', 'Golden Helmet', 'Golden Dart', and 'Blue Shield', they are making incremental improvements in their training. The PLAAF continues to seek engagement with foreign air forces, such as by participating in the Russian-sponsored 'Aviadarts' air-to-ground competition, Shaheen Series with the PAF, and the 'Falcon Strike' exercise with the Royal Thai Air Force, to gain exposure to foreign operational concepts and tactics. The effect of these training initiatives on operational capabilities remains to be seen. However, the PLAAF is beginning to make fundamental changes to the training philosophy that are expected to improve the air capabilities in the coming years. A key factor in this development is the PLAAF's use of advanced air combat maneuvering instrumentation (ACMI) systems, which enable the force to train and monitor air combat training in real time and debrief on aircraft maneuvering, weapons employment, jamming, and chaff and flares, etc. The most prominent components of PLAAF unit training in recent years are the dual concepts of "free air combat" and "unbriefed scenarios". PLAAF unit exercises and drills frequently remove safety restrictions, such as the distance between opposing aircraft, and even refrain from sharing the pre-flight plans of opposing forces in an effort to simulate unscripted scenarios. The importance is on seeing "their own mistakes from their opponent's perspective," in order to "improve and rectify their own weaknesses before the enemy does." The PLAAF has also made advances in joint training. The earlier Army-only exercises now incorporate the PLAAF in a true joint environment.

The efficient integration of different generations of aircraft within the PLAAF remains a work in progress. Pilots are increasingly being deployed to units and airfields other than their home units in an effort to hone skills in reorienting to unfamiliar locations. Concepts involving joint command and control across different branches within the PLAAF as well as across the army, navy, and air force are now gaining importance.

Airborne Assault Brigades

The PLAAF Airborne Corps operates directly under the PLAAF headquarters. It was reorganized and renamed from the 15th Airborne Corps in May 2017, and now comprises six airborne brigades and a special operations brigade.[25] The PLAAF Airborne Corps is China's primary strategic airborne unit and part of the newly formed rapid reaction units of the Chinese military which is primarily designated for airborne and special operation missions. Its role is similar to that of the U.S. Army's 82nd Airborne Division.

Only around three of the seven airborne brigades can deploy to any part of China within 48 hours due to limited airlift capability. The PLAAF Airborne Corps has been elevated to the status of a strategic force. Doctrinal modernization allows the PLAAF Airborne Corps to act as a principal force employed for independent campaign missions in future wars. The airborne troops are envisaged for pre-emptive attacks on the enemy's key military targets in the rear in order to paralyze or disrupt preparation for an offensive. This kind of large-scale mission cannot be conducted without having total control in the air. Also, a single-lift capability of 50,000 men is required for this type of mission. Currently, the PLAAF can lift one division of 11,000 men with light tanks and self-propelled artillery.[26]

PLAAF Airborne Corps headquarters is in Xiaogan, north of Wuhan in Hubei.[27] The earlier airborne divisions were 43rd Division at Kaaifeng, at Henan (127th and 128th Airborne Infantry and 129th Airborne Artillery Regiments), and the 44th and 45th Divisions also in the Wuhan area. Location of new brigades is likely to be similar. More and more focus will be placed on helicopter assaults as opposed to traditional parachute drops. In times of war, the PLAAF Airborne Corps can also utilize transport aircraft such as the Y-20, Y-9, and Y-8, and many smaller utility transport aircraft. During a number of exercises, the PLAAF Airborne Corps has demonstrated it can move a regiment-plus of paratroopers with light armoured vehicles to anywhere within China in less than 24 hours.

The PLAAF Airborne Corps' weapons include self-propelled mortars, jeeps with recoilless rifles or anti-tank guided missiles (ATGM), and self-propelled anti-tank guns. New lightweight high-mobility vehicles have also entered service. Paratroopers are outfitted with portable GPS systems, night-vision goggles, radios and other high-tech equipment.

The Airborne Brigades are further divided into battalions and companies or batteries. In 2015, Chinese paratroopers took first place in the International Army Games held in Russia. The Chinese airborne troops also performed excellently in shooting with small arms, and parachuting. They were among the top in subsequent years too.

PLAAF Vision 2035

China's White Paper on defence was released in July 2019,[28] updating the 2011 document. It states that "Internal security issues" are non-negotiable for China, and any external interference would not be tolerated. China is heading towards its stated goal of PLA informationisation by 2020, modernisation by 2035 and a world class armed force by 2050. China views East and Southeast Asia as its own sphere of influence and aims to keep the USA out of it.

The PLAAF has 62 Regts/Bdes of fighter/bomber aircraft. They will one day stabilise this figure at 80 Regts/Bdes, with all being 4th generation plus aircraft, and at least 20 per cent 5th generation ones. More than the size of the Chinese Air Force, the crucial strength of the PLAAF lies in its establishment of a dense air defence network. There has also been significant progress in the research and development of sixth-generation fighter aircraft within the PLAAF. This sixth-generation aircraft will have across the board new systems and are expected to have artificial intelligence integrated systems and will fly in conjunction with drones. The aircraft is expected to be inducted by 2035. The PLAAF is placing high importance on technology in the conflicts of the future.

China is likely to continue to invest heavily in J-20 stealth fighters and on the qualitative development in avionics and futuristic jet engines. China's biggest strides are coming in air-to-air missiles. The PLAAF believes that with a one or two million dollars weapon, they could destroy a $150 million aircraft.[29] China's latest missile, the PL-15, is claimed to have a greater range than the latest American long range air-to-air missile (160 km) or the European Meteor. The PL-15 also supports an active electronically-scanned array radar that makes evasion difficult even for the most agile fighter jets. Russia has yet to succeed in equipping its own missiles with this technology and the U.S. Air Force has called on Congress to fund a response. Another variant PL-21 would strike slow-moving airborne warning and control systems, from as far away as 400 km. China's new PL-10 missile is comparable to the best "fire-and-forget" dogfight equivalents, and claimed to be much more advanced than the AIM-9 with a range of 60 km. However, none of the Chinese weapons have ever been tested in combat. The Chinese are also known to overstate claims.

Combat modelling by the Rand Corporation think-tank found that China in 2017, for the first time, because of new aircraft, combined with the latest air-to-air, Cruise, anti-ship missiles and Russian S-400 air-defence systems had achieved parity with the USA in air superiority for any conflict close to its mainland, including over Taiwan.[30] Another crucial strength of the PLAAF lies in its dense air defence network. China is also pushing ahead with research and development of sixth-generation fighter aircraft.

Inferences

While Chinese modernization concentrates on technology, challenges continue to face the PLA in the areas of jointness and efficiency. Combined arms warfare remains a weak area. Command and Control (C2) is an issue. The doctrine stresses decentralization, whereas the system is still highly centralised. Narrow individual service interests reduce effective coordination and cooperation in joint operations. Primacy of the Army officers assigned to critical command billets constrains effective coordination and synchronisation in joint operations. The reorganisation of the PLA into five theatre commands was meant to address this issue. More PLAN and PLAAF officers are reportedly planned to be given higher appointments, including command of the theatres.

China's military modernization is dovetailing into Beijing's increasingly assertive foreign policy with new military missions and goals. Military exchange programs are increasing interoperability with the BRI countries, and those that field Chinese weapons systems. China has modernized its forces to win informationised wars by emphasis on network centricity of all weapons systems at sea, air and land connected in real time. The thrust of the PLA's modernization is for non-contact wars.

Introduction of advanced technologies such as ASAT weapons to target space assets and space operations and cyberspace are clear force multipliers. All systems and processes will be driven by Artificial Intelligence (AI), and China is working to go well ahead of the USA.

Such rapid modernisation of the PLAAF has implications for the region. Its target to surpass the USAF would mean it could leave the IAF considerably behind. Since for some foreseeable future, the IAF will not be able to match in class and numbers, there is a need to build deterrence through force multipliers. While building air defence capability, the IAF must also have a good offensive strike capability, both of aircraft and air/surface launched missiles, so as to be able to deter a Chinese misadventure.

REFERENCES

1. PTI Beijing, At $177.6 bn, China's 2019 defence budget is over 3 times that of India, *Business Standard*, March 05, 2019, https://www.business-standard.com/article/pti-stories/china-hikes-defence-budget-by-7-5-per-cent-to-usd-177-61-billion-119030500095_1.html
2. Liu Hui, Graphics: Reform in China's national defence and armed forces, CGTN, 26 Dec 2019, https://news.cgtn.com/news/2019-12-25/Graphics-Reform-in-China-s-national-defense-and-armed-forces-MGRXIeWBZm/index.html
3. Ibid.
4. Anthony H. Cordesman, China's New 2019 Defence White Paper, *Centre for Strategic and International Studies (CSIS)*, July 24, 2019, https://www.csis.org/analysis/chinas-new-2019-defense-white-paper
5. Marcus Clay, Ph.D, "Flying Twice" No More: Assessing the Aerospace Component of the PLA Military Parade, Air University, October 30, 2019, https://www.airuniversity.af.edu/Portals/10/CASI/documents/2019_Oct_PRC_Parade_Aerospace_Power_Recap.pdf?ver=2019-11-01-114023-900
6. A.K. Sachdev, Chinese Missiles: Winning the Limited War, IDSA, https://www.idsa-india.org/an-jun-600.html
7. Ibid.
8. Roger Cliff, John Fei, Jeff Hagen, Elizabeth Hague, Eric Heginbotham, and John Stillion, *Shaking the Heavens and Splitting the Earth*, Chinese Air Force Employment Concepts in the 21st Century, RAND Corporation, 2011, https://www.rand.org/content/dam/rand/pubs/monographs/2011/RAND_MG915.pdf
9. USNI News, Document: China's Military Strategy, May 26, 2015, https://news.usni.org/2015/05/26/document-chinas-military-strategy
10. Defence Intelligence Agency Report 2019, China Military Power - Modernizing a Force to Fight and Win, Appendix C, p. 83, https://www.dia.mil/Portals/27/Documents/News/Military%20Power%20Publications/China_Military_Power_FINAL_5MB_20190103.pdf
11. Lawrence Sid Trevethan, Brigadisation of PLAAF, China Aerospace Studies Institute, August 14, 2018, https://www.airuniversity.af.edu/Portals/10/CASI/Books/Brigadization_of_the_PLA_Air_Force.pdf?ver=2018-08-14-101214-150

12. Cristina L. Garafola, Liberation Army Reforms and Their Ramifications, The RAND Blog, September 23, 2016, https://www.rand.org/blog/2016/09/pla-reforms-and-their-ramifications.html
13. The Military Balance, Chapter Five: Russia and Eurasia,120:1,166-219, DOI: 10.1080/04597222.2020.1707966, p. 264, February 13, 2020, https://doi.org/10.1080/04597222.2020.1707966, https://www.tandfonline.com/doi/abs/10.1080/04597222.2020.1707966
14. Ibid.
15. Arjun Subramaniam, Closing the gap: A doctrinal & capability appraisal of the IAF & the PLAAF, ORF, May 02, 2019, https://www.orfonline.org/expert-speak/closing-gap-doctrinal-capability-appraisal-iaf-plaaf/
16. Ibid.
17. Ibid.
18. Ibid.
19. Roger Cliff, John Fei, Jeff Hagen, Elizabeth Hague, Eric Heginbotham and John Stillion, RAND Corporation (2011), Book Title: *Shaking the Heavens and Splitting the Earth*, Chapter: The Evolution of Chinese Air Force Doctrine, https://www.jstor.org/stable/pdf/10.7249/mg915af.10.pdf
20. Ibid.
21. The People's Liberation Army in the Information Age, Volume 145, Centre for Asia-Pacific Policy (Rand Corporation), p. 162, https://books.google.co.in/books?id =BR7 u98r H5MAC&pg=PA162&lpg=PA162&dq=%E2%80%9Carms+may+be+negatively+ affected+ by+ harsh+ conditions,+ and+ it+is+difficult+ to +organize +logistics +to+sustain+the+war+effort+for+long.%E2%80%9D&source =bl&ots =MQtXfQiVF4&sig =ACfU3U3JCmM-LyaO05UPmowE7I0YEazJRw&hl=en&sa=X&ved=2ahUK EwijlN7 w96LmAhXHzjgGHS ihA6kQ6AEwAHoECAIQAQ#v =one page &q= %E2%80%9Carms%20may%20be%20negatively%20affected%20by%20 harsh %20 conditi ons%2C%20and%20it%20is% 20difficult %20to %20 organize% 20logistics%20to%20sustain%20the%20war%20effort%20for%20long.%E2% 80%9 D&f =false
22. A.K. Sachdev, Chinese Missiles: Winning the Limited War, IDSA, https://www.idsa-india.org/an-jun-600.html
23. Lt Gen (Dr) V.K. Saxena, Growing Muscle of PLAAF, Vivekananda International Foundation Occasional Paper – March 2016, p. 9, https://www.vifindia.org/sites/default/files/growing-muscle-of-plaaf.pdf
24. Lyle J. Morris and Eric Heginbotham, Rand Corporation, From Theory to Practice, PLAAF Aviation Training at Operational Units, 2016, https://www.rand.org/pubs/research_reports/RR1415.html
25. Sudhi Ranjan Sen, *India Today*, China upgrades air capability in Tibet, monitors Indian exercise, May 22, 2018, https://www.indiatoday.in/india/story/china-upgrades-air-capability-in-tibet-monitors-indian-exercise-1239082-2018-05-22
26. Ying Yu Lin, *The Diplomat*, What Are China's Plans for its Airborne Corps? April 20, 2019, https://thediplomat.com/2019/04/what-are-chinas-plans-for-military-aviation/
27. Global Security Org, PLA Airborne Corps - Capabilities, https://www.globalsecurity.org/military/world/china/15abn-corps-capability.htm
28. Anthony H. Cordesman, China's New 2019 Defence White Paper, Centre for Strategic

and International Studies, July 24, 2019, https://www.csis.org/analysis/chinas-new-2019-defense-white-paper

29. Marc Champion, Chinese missiles are transforming balance of power in the skies, *Economic Times*, May 8, 2018, https://economictimes.indiatimes.com/news/defence/chinese-missiles-are-transforming-balance-of-power-in-the-skies/articleshow/64073472.cms?from=mdr

30. Marc Champion, *Live Mint*, Chinese air-to-air missiles are transforming balance of power in the skies, May 08, 2018, https://www.livemint.com/Politics/9UtKiySJfZYIm0t3y4vDIO/Chinese-airtoair-missiles-are-transforming-balance-of-powe.html

6

Space:
China the Emerging Power

"The Earth is the cradle of humanity, but mankind cannot stay in the cradle forever."

—Konstantin Tsiolkovsky

"We will build new ships to carry man forward into the universe, to gain a new foothold on the moon and to prepare for new journeys to the worlds beyond our own."

—George W. Bush

In June 2019, China became the third country to successfully launch an orbital space mission from a sea-based platform. A Long March 11 rocket carrying seven satellites blasted off from a large semi-submerged barge in the Yellow Sea.[1] The flexibility afforded by sea launches will allow China to provide better commercial launch services for countries along the Belt and Road network. Interestingly, all of China's launch sites are otherwise located inland. Nearly 50 years after China launched its first satellite, the 173-kg Dong Fang Hong into orbit in April 1970, Beijing is strengthening its foothold in space. China's was the heaviest first satellite placed into orbit by a nation, exceeding the combined masses of the first satellites of the other four previous countries.[2] China is already building a space station and is getting all set for its Mars Lander mission.[3] China has developed the powerful Long March 5 rocket to transport the probe to Mars in 2020.

Elon Musk says he's a big fan of what China is doing in space. Musk is Chief Executive Officer (CEO) of the space exploration company, SpaceX, that faces growing competition from China which is pumping huge money into space.[4] The first Chinese manned spaceflight was in 2003. In January 2007, China became the first Asian military-space power to send an anti-satellite missile into orbit, and destroy an aging Chinese weather satellite.[5] Anti-satellite technologies to destroy or disable space-based assets are a critical part of the Chinese space program. These include land-based missiles, experimental lasers, and signal jammers. China has successfully performed a soft landing of a rover on the moon, including on the only ones to land on the dark side.[6] China has long-term ambitions to exploit Earth-Moon space for industrial development. China plans to bring a habitable space station Tiangong2online by 2022[7] and put Chinese astronauts on the moon in the mid-2020s. They also have a Mars Lander mission coming up. The Chinese space program is linked to the nation's efforts at developing advanced military technology. China launched the Dark Matter Particle Explorer (DAMPE), the most capable dark matter explorer to date in 2015,[8] and the world's first Quantum Experiments at Space Scale 'QUESS' satellite in 2016.[9] China is averaging 20 space missions a year.[10] As per estimates, China has over 500 ballistic missiles including 100 Inter Continental Ballistic Missiles (ICBM), 25 per cent of which are submarine based, and some with Multiple Independent Re-entry Vehicle (MIRV) warheads, with ranges beyond 13,000 kilometres.

China's Space Program: Evolution

China's space program officially began in 1956 with the creation of the Fifth Academy of the Ministry of National Defence, presently the China Aerospace Science and Technology Corporation, which conducts most of China's space research and development. During a speech to the Communist Party in 1958, Mao Zedong, the founder of modern China, declared that Beijing should also launch an artificial satellite into space, following successful launches by the Soviet Union and the USA.

Mao wanted China to gain equal, if not better, status among the world's powers who, he felt, did not respect him. His goal would be achieved in phases. The construction of China's first missile test base, code-named Base 20, started in April 1958 and it entered service on October 20 of the same year. The first Chinese missile was built in October 1958 as a reverse-engineered copy of the Soviet R-2 short-range-ballistic missile (SRBM), itself an upgraded version of a German V-2 rocket.[11] Its range was 590 km, weighed 20.5 tons and was propelled with liquid oxygen and alcohol. China's first ever T-7 sounding rocket was successfully launched from the Nanhui launch site on February 19, 1960. During the cordial Sino-Soviet relations of the 1950s, the USSR engaged in a cooperative technology transfer program with the PRC under which it trained Chinese students and provided the fledgling program support. After the friendly relationship between the two countries turned to confrontation, all Soviet technological assistance was abruptly withdrawn after the 1960 Sino-Soviet split.

The first successful launch of a Chinese SRBM missile was conducted on November 5, 1960. The missile was now designated DF-1. The first DF-2 MRBM was successfully tested on June 29, 1964. It would enter service by the end of 1966. China started to develop the DF-5 Inter Continental Ballistic Missile (ICBM) program in August 1965. It was designed to carry a single nuclear warhead with a maximum range of 12,000 km.[12] In November 1966, it was decided to build a second ballistic missile test site, the Northern Missile test Site in Shanxi Province. On October 27, 1966, a nuclear-tipped DF-2A missile was launched from Jiuquan, and a 20 kiloton-yield nuclear warhead exploded at a height of 569 metres over the target in Lap Nor or Base 21.[13] In March 1967, development started on the JL-1 submarine-launched ballistic missile to be on board the Type 092 ballistic missile submarine which was being developed in parallel.

DF-5B Inter Continental Ballistic Missile.
Picture Source: Military Watch Magazine

Space Organisational Structure

Initially, the space program of the PRC was organized under the PLA Second Artillery Corps. In the 1990s, the PRC reorganized the space program as part of a general reorganisation of the defence industry to make it resemble Western defence procurement. The State Council's State Administration for Science, Technology, and Industry for National Defence (SASTIND) is the primary civil organization that coordinates and manages China's space activities, including allocating space research and development funds. It works closely with the PLA. The China National Space Administration is an agency within the SASTIND. China has a large number universities doing research on Aerospace and Astronautics. The Chinese space program has close links with these universities. The PRC operates four satellite launch centres,[14] one each at Jiuquan in Gobi desert, close to Mongolia; Xichang, just south of Chengdu and not too far from Arunachal Pradesh; Taiyuan, midway between Beijing and Xi'an; and Wenchang on the Hainan Island. The main monitoring and control centres are at Beijing and Xi'an. China also has a new integrated land-based space monitoring and control network stations, forming a large triangle with Kashi in the north-west of China, Jiamusi in the north-east and Sanya in the south. The overseas tracking stations are at Karachi in Pakistan, Malindi in Kenya, Swakopmund in Namibia, and Neuquen province in Argentina. They also share space tracking facilities with France, Brazil, Sweden and Australia.

Chinese Launch Rockets Program

It is worth having an overview of Chinese space rockets. A CZ-2F rocket launched the first Chinese astronaut, Yang Liwei, into space in October 2003.[15] New rockets are being designed to be more environmentally friendly. While previous Chinese launchers often used toxic hypergolic propellants, the new designs use kerosene or hydrogen for the first and second stage engines,[16] The new rocket's CZ-6 small payload launcher first flew in 2015, the CZ-7 medium lift rocket and the CZ-5 heavy lift rocket both had maiden flights in 2016. Developed from the CZ-2E launcher, the CZ-2F is used to support the Chinese human spaceflight effort. It was also used for rendezvous and docking with the Tiangong 1 space laboratory. The CZ-2E (A) intended for launch of Chinese space station modules has a payload capacity up to 14 tons in LEO and 9000 (kN) liftoff thrust developed by 12 rocket engines, with enlarged fairing to accommodate large spacecraft. CZ-3 rockets are a series of three-stage geosynchronous satellite launchers. The CZ-5 is a Chinese heavy lift space launch system. The core stage uses liquid hydrogen (LH) and liquid oxygen (LOX) as propellants. The re-startable second stage burns LH and LOX. An optional third stage is available for some missions. The maiden flight of a CZ-5 rocket was successfully flown in November 2016. The second CZ-5, launched in 2017, failed to place its payload into the correct orbit. The CZ-5 is meant for the Chinese space station program and lunar program.[17] The CZ-7 is a medium lift rocket that first flew in 2016. The rocket is designed to include up to four strap-on boosters. The CZ-7 is expected to replace the CZ-2F/G rockets supporting human spaceflight. CZ-9 will be a super heavy launch vehicle that is currently in study. It is planned for a maximum payload capacity of 140,000 kg to LEO, 50,000 kg to Lunar Transfer Orbit or 44,000 kg to Mars. CZ-11 was first launched in 2015 from a canister attached to a mobile transporter/erector in a manner similar to an ICBM. It is meant for a quick-reaction orbital launcher. The 58-tonne launch mass and a 120 tonne liftoff thrust CZ-11 may be able to lift 700 kg to sun synchronous orbit.[18]

China is improving its space launch capabilities to ensure it has an independent, reliable means to access space and compete in the international space launch market. Modular SLVs allow China to tailor an SLV to a specific configuration and overall cost savings for launch campaigns.

China's Current Space Launch Capabilities[19]

The immense technical expertise and financial resources needed to develop rockets powerful enough to successfully deliver payloads into orbit precludes most countries from developing indigenous launch capabilities. This leaves the vast majority of countries reliant on foreign launch vehicles for their missions in space. Even the USA relies on foreign-built rockets for launching some of its payloads. China, however, bucks this trend. Over 96 per cent of the Chinese satellites currently in orbit were launched aboard Chinese-made rockets. The Long March series of rockets, which includes several variants with different capabilities, is at the core of China's space launch capabilities. Long March rockets are designed and developed by the state-owned defence industry giant, China Aerospace Science and Technology Corporation (CASC). A CASC subsidiary, the China Academy of Launch Vehicle Technology, has produced nearly all 15 variants of the Long March series, including the Long March 1 (CZ-1) that carried China's first satellite into orbit in 1970. Long March rockets have carried out almost all Chinese space launches. By October 2019, the Long March series had completed 299 successful launches, and faced seven partial failures and nine failures. This gives the series a success rate of 94.4 per cent, which compares closely with key launch vehicles of other countries. The Delta series, which currently operates as an integral component of US military efforts in space, has a 95.5 per cent success rate. Russia's Soyuz rocket family has completed more launches than any other rocket series in history and enjoys a success rate of 97.2 per cent.

Chinese Space Launch Vehicles.
Picture Source: Space.com

China's most successful Long March variants, the CZ-2F, has had all its 13 launches successful. The CZ-2F was designed to launch the Shenzhou spacecraft, which has supported all six of China's manned missions. The USA ended its space shuttle program in 2011, leaving it dependent on Russian rockets for manned missions. However, NASA's new Space Launch System is being developed for a manned US mission to the moon, planned for 2024.

Long March Launches

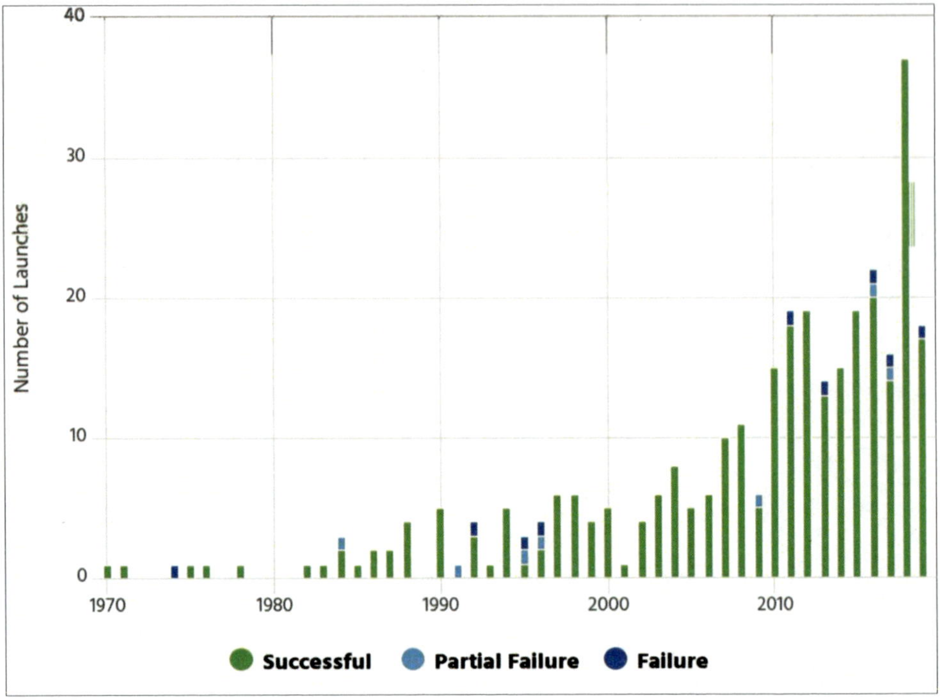

Source: CSIS China Power Projects[20]

Chinese Crewed Space Program

As the space race between the two superpowers reached its climax with the conquest of the Moon, Mao decided on July 14, 1967 that the PRC should start China's own crewed space program. China's first spacecraft designed for human occupancy was named Shuguang-1. It was decided to construct a new space centre in the mountainous region of Xichang in the Sichuan province, far away from the Soviet border. In August 1969, the development of China's first heavy-lift satellite launch vehicle (SLV), the Feng Bao 1 (FB-1) was the all-liquid two-stage launcher derived from the DF-5 ICBM.[21] The first DF-4 liquid-propellant with two-stage, single-warhead IRBM was tested with success on January 30, 1970. The addition of a second-stage allowed the missile to increase its range to over 4,750 km. China succeeded in its second satellite launch attempt on April 24, 1970. It was the heaviest first satellite placed into orbit by a nation, exceeding the combined masses of the first satellites of the other four previous countries. The third stage of the CZ-1 was specially equipped with a 40 m^2 solar reflector deployed by the centrifugal force developed by the spin up orbital insertion solid propellant stage. The PRC's second satellite, the 221-kg ShiJian-1 (SJ-1) was launched with the last of the CZ-1 SLVs on March 3, 1971. The first crewed space program known as Project 714, was officially adopted in April 1971 but the program would soon be cancelled due to political turmoil.

The CZ-2C launcher successfully launched a recoverable satellite into orbit on November 26, 1975. After Mao died on September 9, 1976, his rival, Deng Xiaoping, slowly re-emerged as China's new leader in 1978. He had a re-look at the entire program. Further development of the Long March rocket series allowed the PRC to initiate a commercial launch program in 1985, which has since launched over 30 foreign satellites, primarily European and Asian. The Ministry of Aerospace Industry was founded on July 5, 1988. On September 15, 1988, a JL-1 SLBM was launched from a submarine. The maximum range of the SLBM was 2150 km.[22] Along Deng's policy of capitalist reforms in the Chinese economy, Chinese culture also changed. Therefore, names used in the space program, previously chosen from the revolutionary history of the PRC, were soon replaced with mystical-religious ones. Thus, the Long March carrier rockets were renamed Divine Arrow, spacecraft as Divine Vessel, land-based high-power laser as Divine Light.

In 1992, authorization and funding was given for the first phase of Project 921, which was a plan to launch a crewed spacecraft. In June 1993, the China Aerospace Industry Corporation (National Space Bureau) was founded in Beijing. The Shenzhou program had four unmanned test flights and two crewed missions. On the 50th anniversary of the PRC's founding, China launched the Shenzhou 1 spacecraft on November 20, 1999 and recovered it after a flight of 21 hours. On January 9, 2001, Shenzhou 2 carried test animals. Shenzhou 3 and 4 were launched in 2002, carrying test dummies. Shenzhou 5 was China's first crewed mission in space on October 15, 2003, which carried Yang Liwei in orbit for 21 hours and made China the third nation to launch a human into orbit.[23] Shenzhou 6 followed two years later ending the first phase of Project 921. The missions used Long March 2F rockets from the Jiuquin Launch Centre.

Chinese Space Laboratory

The second phase of Project 921 started with Shenzhou 7, China's first spacewalk mission. Then, two crewed missions were planned to the first

Chinese space laboratory. The PRC initially designed the Shenzhou spacecraft with docking technologies imported from Russia, therefore compatible with the International Space Station (ISS). On September 29, 2011, China launched Tiangong 1. Tiangong means 'Heavenly Palace' in Chinese. This target module was intended to be the first step to testing the technology required for a planned space station. On October 31, 2011, a Long March 2F rocket lifted the Shenzhou 8 unmanned spacecraft which docked twice with the Tiangong 1 module.[24] The Shenzhou 9 craft took off on 16 June 2012 with a crew of three 'taikonauts'. It successfully docked with the Tiangong-1 laboratory on 18 June 2012, marking China's first crewed spacecraft docking.[25] Another crewed mission, Shenzhou 10, was launched on 11 June 2013 and returned on 26 June 2013. The Tiangong 1 target module served as both a manned laboratory and an experimental test-bed to demonstrate orbital rendezvous and docking capabilities during its two years of active operational life. It de-orbited on 2 April 2018, much longer than originally planned. A second space lab, Tiangong 2, was launched on 15 September 2016. Shenzhou 11 launched and rendezvoused with Tiangong 2 in October 2016,[26] with an unconfirmed further mission Shenzhou 12 in the future. The Tiangong 2 has the POLAR gamma ray burst detector, a space-Earth quantum key distribution and laser communications experiment, to be used in conjunction with the Mozi 'Quantum Science Satellite', a liquid bridge thermo-capillary convection experiment and a space material experiment.[27] Also included is a stereoscopic microwave altimeter, a space plant growth experiment, and a multi-angle wide-spectral imager and multi-spectral limb imaging spectrometer. Onboard TG-2 there is also be the world's first-ever in-space cold atomic fountain clock.

Tiangong 2 Space Lab.
Picture Source: spacenews.com

A little over a year after TG-1 crashed down over the Pacific Ocean, China de-orbited the Tiangong-2 in a controlled fashion over the Pacific, where there wasn't any risk of it hitting people.[28] China had proposed a third station, the Tiangong-3, but it was never constructed or launched. China was planning to launch in April 2020 the country's next space station. A Long March 5B rocket is to carry a "trial version" of China's new spaceship, which is designed to carry crews of up to six people.[29] Previous Chinese spaceships carried up to three taikonauts. While the April mission was to carry neither people nor pieces of the future space station, it is expected to prepare Chinese officials for space station construction, the report added. The same may now get delayed due to the COVID-19 virus spread.

The permanent future space station called 'Chinese large modular space station' is planned to be assembled by 2022.[30] The design of Tianzhou, an automated cargo spacecraft intended to resupply the Chinese large modular space station, is based on Tiangong-1. The Chinese Permanent Space Station with permanent occupation will have a robotic version of the Shenzhou spacecraft for resupply and return cargo back to Earth. The Tianzhou robotic cargo resupply vessel is not meant for re-entry, but usable for garbage disposal.

Chinese Lunar Exploration Program[31]

The Chinese Lunar Exploration Program is divided into four main operational phases, with each mission serving as a technology demonstrator in preparation for future missions. The first phase of two lunar orbiters is complete. Chang'e 1, launched on 24 October 2007 scanned the entire Moon in unprecedented detail, generating a high definition 3D map that would provide a reference for future soft landings. The probe also mapped the abundance and distribution of various chemical elements on the lunar surface as part of an evaluation of potentially useful resources. Chang'e 2 launched on 01 October 2010, reached the Moon in less than five days, compared to 12 days for Chang'e 1, and mapped the Moon in even greater detail. It then left lunar orbit and returned. The second phase which is still ongoing incorporates spacecraft capable of soft-landing on the Moon and deploying lunar rovers. Chang'e 3 launched on 02 December 2013 landed on the Moon on 14 December 2013. It carried a 140-kg lunar rover 'Yutu', which was designed to explore an area of 3 square kilometres for three months, and cover many experiments. Chang'e 4 was launched on 07 December 2018. It landed on the South Pole (Aitken Basin), on the far side of the Moon on January 3, 2019, and deployed the Yutu-2 rover. The rover was able to transmit data back to Earth despite the lack of radio frequencies on the far side, via a dedicated satellite sent earlier to orbit the moon. The landing and data transmission is considered a landmark achievement for human space exploration. The third phase was for sample return missions. Chang'e 5T1 (Test) was launched on 23 October 2014, to test lunar return spacecraft. Chang'e 5 is expected to be launched in 2020, aboard a Long March 5 rocket, and carry a Lander capable of collecting up to 2 kilograms of lunar samples and returning with them to Earth. Phase Four is for development of a robotic research station near the Moon's south pole. Chang'e 6, 7, 8 are expected to be launched from 2023 onwards, to ultimately look for useable natural resources and may 3D-print a structure using regolith. A small scaled ecosystem experiment is also planned. It will test technology necessary for the construction of a lunar science base.

China's Deep Space Exploration Plan

Chinese researchers proposed a deep space exploration roadmap to explore Mars, an asteroid, Jupiter, and further targets, within the 2020–2030 timeframe. The Mars Global Remote Sensing Orbiter and Small Rover (HX-1) was planned for launch in July–August 2020 with the landing on Mars in February 2021.[32] The mission includes an orbiter, a Lander, and a rover. An Asteroid Exploration Mission (ZhengHe) is proposed for launch around 2022–2024. Mission goals include asteroid flyby observations and Mars Sample Return Mission (HX-2), proposed for launch around 2028–2030. Mission goals include in-situ topography and soil composition analysis, deep interior investigations to probe the planet's origins and geologic evolution, and sample return. The Jupiter System Exploration Mission is proposed for launch around 2029–2030, and arrive at Jupiter around 2036.[33] Mission goals include orbital exploration of Jupiter and its four largest moons, study of the magneto-hydro-dynamics, and investigation of the internal composition of Jupiter's atmosphere and moons. A mission to Uranus for implementation after 2030, with a probe arriving in the 2040s. It is presumed to be part of a future planetary flyby phase of exploration, and would study the solar wind and interplanetary field as well. Mars crewed phase is in 2040-2060.[34] Chinese interest in space-based solar power began in the 1990s. Once completed, the solar station, with a capacity of 100MW, could span at least one square kilometre, dwarfing the International Space Station and becoming the biggest man-made object in space. China plans to accomplish a 200-tonne megawatt-level space-based solar power station by 2035, according to the China Academy of Space Technology.[35]

BeiDou Navigation Satellite System[36]

The BeiDou Chinese satellite navigation system was built in phases. The first, BeiDou-1, consisted of three satellites which since 2000 had offered limited coverage and navigation services, mainly for users in China and neighbouring regions. Beidou-1 was decommissioned in 2012. The second system, called the BeiDou Navigation Satellite System (BDS) and also known as COMPASS or BeiDou-2, became operational in December 2011 with a partial constellation of 10 satellites in orbit. Since December 2012, it has been offering services to customers in the Asia-Pacific region. In 2015, China launched the third generation BeiDou-3 system with global coverage constellation. The first BDS-3 satellite was launched on 30 March 2015. On 27 December 2018, the BeiDou Navigation Satellite System started providing global services with around 15 satellites. BeiDou-3 consists of 35 satellites, and provides an alternative global navigation satellite system to the U.S. Global Positioning System (GPS), Russian GLONASS, and European Galileo systems. The Chinese claim it will be more accurate with millimetre-level accuracy with post-processing. According to a *China Daily*, report of 2015, fifteen years after the satellite system was launched, it was generating a turnover of $ 31.5 billion per annum for major Chinese companies such as China Aerospace Science and Industry Corporation, AutoNavi Holdings Ltd and China North Industries Group Corporation. In March 2020, China launched a satellite among the last few to take it closer to completing the third phase of the BDS. Project. Director Ran Chengqi reported that two more satellite launches will complete the 3rd generation and global BeiDou system by June 2020. The same has actually been achieved. China's navigation system already boasts the largest fleet of satellites. The BeiDou-3 constellation has over taken the U.S. GPS's 31 operational satellites. In comparison, GlOSNASS has 26 and Galileo 26 satellites; however, not all are currently operational. Outnumbering the three other satellite providers, China is well on its way to becoming the world leader in technology and space. The BDS will be used for positioning, timing and wide-area differential and short message communications. It can also be used for public security, forestry, fishing, search and rescue, etc. The satellites were also used to combat the corona virus epidemic in China.[37]

Creating a rival to GPS, China's goal is to become the world leader in space. "The main advantage of having your own system is security of access, in the sense that you are not relying on another country to provide it. The USA could deny users access over certain areas, for example, in times of conflict," Alexandra Stickings of the Royal United Services Institute for Defence and Security Studies said.[38] This new "Space Silk Road" has progressed rapidly. The rate at which China has produced this technology reveals the ambition and technical prowess of China. Global coverage of BeiDou was reached two years ahead of schedule. The last 19 BDS-3 satellites were launched within as little as 17 days between some launches, setting a new record in the history of the development of the world's satellite navigation systems.

Military Applications of BDS

BeiDou has already shipped over 70 million systems to more than 90 countries. These systems include microchips and modules, which could give China easier access.[39] The BDS will also play an important military role including reconnaissance and short messaging platform, both of which are not in GPS. It is claimed to have greater than military accuracy and holds the potential to "deny or degrade the signal received by other users of the system," according to Stickings.[40] BeiDou satellites, in the event of a conflict, will enable the Chinese military to identify, track and strike U.S. ships, increasing their tracking ability by 100 to 1,000 times. It will also support precision-guided missiles, smart bombs, navigation and operation of ships, other vehicles and troops. China's actions are continuing to fuel a space race. Meanwhile, Russia and China have signed "the Agreement on Peaceful Use of BeiDou and GLONASS", envisaging large-scale cooperation in the satellite navigation field.

Space Program PLA Linkages

It can be seen that China has devoted significant economic and political resources to improving all aspects of its space program, from improving military space applications to developing human spaceflight and lunar exploration programs. China's space program is moving with a sense of purpose. Beijing is working to build China into a space power in all respects. Its rapidly growing space program is second only to the USA. It is a part of President Xi Jinping's plan to establish a powerful and prosperous China. The space program supports the military modernisation efforts. China officially advocates the peaceful use of space, and is pursuing agreements at the United Nations on the non-weaponisation of space. Yet, it continues to improve its counter-space weapons capabilities and has enacted military reforms to better integrate cyberspace, space, and EW into joint military operations.

The PLA routinely incorporates jamming and anti-jamming techniques against multiple communication, radar systems, and GPS satellite systems in exercises. China continues to develop jammers dedicated to targeting SAR aboard military reconnaissance platforms, including LEO satellites.[41] Additionally, China is developing jammers to target SATCOM over a range of frequency bands, including military protected extremely high frequency communications.

The PLA views space superiority, the ability to control the information sphere, and denying adversaries the same as key components of conducting modern information-centric wars. Space and counter-space operations are now an integral part of PLA campaigns. China has been trying to replicate the US model, and beat them at it. The PLA feels "destroying or capturing satellites and other sensors" would make it difficult for an enemy to use precision guided weapons. Reconnaissance, communications, navigation, and early warning satellites could be among the targets of attacks. The fact that the PLA has historically managed China's space program has a significant military angle. As part of the military reforms announced in 2015, China established the Strategic Support Force (SSF) to integrate cyberspace, space, and EW capabilities into joint military operations. The SSF is the core of China's information warfare force.[42] Chinese military strategy documents also emphasize the growing importance of offensive air, long-distance mobility, and space and cyberspace operations.

Militarisation of Space Program

China is likely pursuing laser weapons to disrupt, degrade, or damage satellites and their sensors and possibly already has a limited capability to employ laser systems against satellite sensors. China is likely to field ground-based laser weapons that can counter low-orbit space-based sensors by 2020, and by the mid-to-late 2020s, it may field higher power systems that extend the threat to the structures of non-optical satellites.[43]

China is developing sophisticated on-orbit capabilities, such as satellite inspection and repair, at least some of which could also function as weapons. The PLA has an operational ground-based ASAT missile intended to target LEO satellites. China has also formed military units that have begun training with ASAT missiles. China probably intends to pursue additional ASAT weapons capable of destroying satellites up to GEO. In 2013, China launched an object into space on a ballistic trajectory with a peak altitude above 30,000 km.

The militarization of China's space program is without doubt. It has serious implications for its adversaries, including India. The high investment is difficult to match. India would need to keep pushing its space program to not let the gap grow too large.

Other Futuristic Space Initiatives

Some of the other Chinese space-related initiatives are a space-based ASAT system comprising small and nano-satellites developed by the Small Satellites Research Institute. The Double Star mission comprised two satellites launched in 2003 and 2004, jointly with the European Space Agency (ESA) to study the Earth's magnetosphere. China has a large number of Earth observation, remote sensing, or reconnaissance satellites. Tianlian 2 will be the next generation data relay satellites system, based on the DFH-4 satellite bus, with two satellites providing up to 85 per cent coverage.[44] More advances in China's BeiDou navigation system with 60-70 satellites. Astrophysics research with the world's largest Solar Space Telescope launched in 2008, and a project of Space Hard X-Ray Modulation telescope. Kuafu mission satellites for space weather forecasts with Canada and ESA is currently held back. China has the deep space tracking network using the world's largest single dish radio antenna of 500 m diameter in Guizhou,[45] and a 3000 km distant VLB radio antenna. There is a plan for Deep-impact style mission to test process of re-directing the direction of an asteroid or comet.

Space Services: Exports

China also exports its satellite technology globally, including its indigenously-developed communications satellites. China intends to provide SATCOM support to users worldwide and has plans to develop at least three new communications constellations. China also intends to use its BeiDou constellation to offer additional services and incentives to countries taking part in the "Belt and Road Initiative". It has provided 15-cm telescopes to Peru, Pakistan, and Iran that are capable of tracking objects in LEO and GEO.[46] All tasking information and subsequent observation data collected is funnelled through the Chinese Academy of Science's National Astronomical Observatory of China.

Security and Proprietary Concerns

The PRC is a member of the United Nations Committee on the Peaceful Uses of Outer Space and a signatory to all other UN treaties and conventions on space, with the exception of the unpopular 1979 Moon Treaty. The United States government has long been resistant to the use of PRC launch services by American industry due to concerns over alleged civilian technology transfer that could have dual-use military applications to countries such as North Korea, Iran, and Syria, and announced an official embargo against the PRC in 2000. Financial retaliatory measures have been taken on many occasions against several Chinese space companies. Due to security concerns, all researchers from the U.S. National Aeronautics and Space Administration (NASA) are prohibited from working with Chinese citizens affiliated with a Chinese state enterprise or entity. In April 2011, NASA was banned from using its funds to host Chinese visitors at NASA facilities.[47] In March 2013, the U.S. Congress passed legislation barring Chinese nationals from entering NASA facilities without a waiver from NASA. The Chinese response to the exclusion policy involved its own space policy of opening up its space station to the outside world, welcoming scientists coming from all countries.

There's every chance that China, as the ultimate disruptor,[48] will do to the commercial space industry what it has already managed with manufacturing: cut costs, boost efficiency and storm into a lead. Launching miniature satellites, which typically weigh less than 220 lb., will become a $ 15 billion industry by 2027, according to Euroconsult, a space industry consulting firm. One firm looking to capitalize on the situation is the Beijing-based start-up i-Space, which is preparing for its third launch in coming weeks. Like most of the $500 million-valued firm's employees, CEO Jingqi Cai came from China's state space industry. She sees no limit to what the Chinese commercial space industry can achieve. "I don't know any country in the world which can do things as fast as in China," says Cai.

China's Space Program: Military Implications

As of May 2018, the Chinese reconnaissance and remote sensing fleet consisted of more than 120 satellites designed to collect data for civil, commercial, or military owners and operators. Reportedly, the PLA owns and operates about half of these systems, most of which could support monitoring, tracking, and targeting.[49] These satellites also allow the PLA to maintain situational awareness of China's regional rivals including India and Japan and potential regional flashpoints like Korea, Taiwan, and the East and South China Seas. China's light, quick response space launch vehicles are capable of expedited launches, and have military implications. These require little launch infrastructure, can relocate by road or rail and can be stored launch-ready for longer periods.

China is already developing super heavy-lift space launch vehicles, for crewed lunar and Mars exploration missions. It is building capability to jam global navigation and communications satellites used for command and control of naval, ground, and air forces, to include manned and unmanned vehicles. Weapons designed to target ISR capabilities would deny the adversary the ability to locate, monitor, track, and target the enemy. Lasers can temporarily or permanently blind imagery satellites and other strategic sensors. China has anti-satellite missiles to shoot down satellites in low Earth orbit. China is believed to be testing on-orbit, space-based anti-satellite technologies and concepts. Physical attacks against ground sites and infrastructure that support space operations can also threaten satellite services. China's cyber capabilities could target space systems and supporting infrastructure, including communications and GPS. China considers both offensive cyber capabilities and electronic warfare as key assets for maintaining military advantage. China intends to field counter-space directed-energy weapons. 3D printing in space could allow China to nearly eliminate the cost of space launch by servicing, repairing, or manufacturing new systems entirely on orbit. China's space program requires a close watch.

REFERENCES

1. Hanneke Weitering, China Aces Its 1st Rocket Launch at Sea, Puts 7 Satellites in Orbit, Space.com, June 05, 2019, https://www.space.com/china-first-sea-rocket-launch-success.html
2. Mathew S. Williams, All You Need to Know About The Chinese Space Program, Interesting Engineering, March 16, 2019, https://interestingengineering.com/all-you-need-to-know-about-the-chinese-space-program
3. Martin Pollard, China completes crucial landing test for first Mars mission in 2020, Reuters, November 14, 2019, https://www.reuters.com/article/us-space-exploration-china-mars/china-completes-crucial-landing-test-for-first-mars-mission-in-2020-idUSKBN1XO0IQ

4. Isobel Asher Hamilton, Elon Musk says he's a big fan of what China is doing in space, Business Insider, July 19,2019, https://www.businessinsider.in/elon-musk-says-hes-a-big-fan-of-what-china-is-doing-in-space/articleshow/70293973.cms
5. Carin Zissis, China's Anti-Satellite Test, Council on Foreign Relations, https://www.cfr.org/backgrounder/chinas-anti-satellite-test
6. Sarah Pruitt, China Makes Historic Landing on 'Dark Side' of the Moon, history.com January 09, 2019, https://www.history.com/news/china-plans-historic-landing-on-dark-side-of-the-moon
7. Elizabeth Howell, China wants to build a new space station. space.com February 22, 2020, https://www.space.com/china-space-station-april-launch-will-prepare.html
8. Charles Q. Choi, China's Cosmic 'Monkey King' Satellite Looks for Dark Matter, space.com, November 29, 2017, https://www.space.com/38937-china-monkey-king-satellite-dark-matter.html
9. China launches world's first quantum science satellite, physicsworld.com August 16, 2016, https://physicsworld.com/a/china-launches-worlds-first-quantum-science-satellite/
10. How is China advancing its space launch capabilities? China power, CSIS.org, https://chinapower.csis.org/china-space-launch/
11. Edited by Air Cmde Jasjit Singh, *China's India War, 1962: Looking Back to See the Future*, Chapter 11, KW Publishers Pvt. Ltd., March 15, 2013.
12. Ibid.
13. Matteo Damiani, Chinese Space Program: a Photographic History, china-underground.com June 07, 2017.
14. Chinese Space Facilities, GlobalSecurity.org, https://www.globalsecurity.org/space/world/china/facility.htm
15. Emma Young and David Chandler, First Chinese astronaut blasts off, *New Scientist*, October 15, 2003, https://www.newscientist.com/article/dn4275-first-chinese-astronaut-blasts-off/#ixzz6HhPNy8dg
16. Chinese Rockets Overview, Historic Spacecraft, Chinese Launch Vehicles, https://historicspacecraft.com/Rockets_Chinese.html
17. How is China advancing its space launch capabilities? China power, CSIS.org, https://chinapower.csis.org/china-space-launch/
18. Space Launch Report, July-September, 2015. http://www.spacelaunchreport.com/slr2015q3.html
19. How is China advancing its space launch capabilities? China power, CSIS.org, https://chinapower.csis.org/china-space-launch/
20. Ibid.
21. Chinese Space Program, Fastogo.in July 05, 2019, https://fastgo.in/webdesign/chinese-space-program/
22. Edited by Air Cmde Jasjit Singh, *China's India War, 1962: Looking Back to See the Future*, Chapter 11, KW Publishers Pvt. Ltd, March 15, 2013.
23. Erik Gregersen, Shenzhou Chinese Spacecraft, *Encyclopaedia Britannica*, January 30, 2020, https://www.britannica.com/technology/Shenzhou
24. Mike Wall, China Succeeds in First Space Docking by 2 Spaceships, Space.com, November 02, 2011, https://www.space.com/13477-china-space-docking-success-shenzhou8-tiangong1.html
25. Jonathan Amos, Science correspondent, Shenzhou-9 docks with Tiangong-1, BBC News, June 18, 2012, https://www.bbc.com/news/science-environment-18481806

26. Tariq Malik, China's Tiangong-2 Space Lab Mission in Pictures, Space.com, September 15, 2016, https://www.space.com/34081-china-tiangong-2-space-lab-in-pictures.html
27. Rui C. Barbosa, China launches Tiangong-2 orbital module, NasaSpaceflight.com September 14, 2016, https://www.nasaspaceflight.com/2016/09/china-launch-tiangong-2-orbital-module/
28. Andrew Liptak, China has deorbited its experimental space station, *The Verge*, July 20, 2019, https://www.theverge.com/2019/7/20/20701831/china-tiangong-2-deorbited-experimental-space-station
29. Elizabeth Howell, China wants to build a new space station. A planned launch in April will set the stage, Space.com, February 22, 2020, https://www.space.com/china-space-station-april-launch-will-prepare.html
30. Ibid.
31. Chinese Lunar Exploration Program, Wikipedia, https://en.wikipedia.org/wiki/Chinese_Lunar_Exploration_Program
32. Mike Wall, Mars Missions Are One Year Away from Launching to the Red Planet in July 2020, Space.com, July 25, 2019, https://www.space.com/4-mars-missions-one-year-from-july-2020-launch.html
33. Khor Eng Lee, Aaron Khor, Book: *China's long march of modernisation : blueprint & road map for the nation's Full Development 2016-2049*. Publisher: Xlibris AU (11 December 2019) Chapter New China A(6) Ascending High Ground.
34. Ajay Lele, Book: *Asian Space Race: Rhetoric or Reality?* Springer, IDSA, p. 89.
35. Xinhua, China to build space-based solar power station by 2035, chinadaily.com.cn, December 02, 2019, https://www.chinadaily.com.cn/a/201912/02/WS5de47aa8a310cf3e3557b515.html
36. BeiDou Navigation Satellite System, Chinese Government Beidou Website http://en.beidou.gov.cn/SYSTEMS/System/
37. Tech2 News Staff, China successfully launches beidou satellite, comes closer to completing third-gen global satellite navigation system, *First Post*, March 11, 2020, https://www.firstpost.com/tech/science/china-successfully-launches-beidou-satellite-comes-closer-to-completing-third-gen-global-satellite-navigation-system-8140611.html
38. Yao Jianing, China's BeiDou Navigation System Goes Global, english.pladaily.com, Source '*The Trumpet*', March 01, 2019 http://english.pladaily.com.cn/view/2019-01/03/content_9394881.htm).
39. Ibid.
40. Ibid.
41. Andrew S. Erikson, China Analysis from Original Reports, New DIA Report - Challenges to Security in Space, February 11, 2019 http://www.andrewerickson.com/2019/02/new-dia-report-challenges-to-security-in-space-offers-copious-china-related-information/
42. John Costello and Joe McReynolds, China's Strategic Support Force: A Force for a New Era, Executive Summary, Centre for the Study of Chinese Military Affairs Institute for National Strategic Studies National Defence University, October 2018, https://ndupress.ndu.edu/Portals/68/Documents/stratperspective/china/china-perspectives_13.pdf
43. Michael Davis, Laser Guns in Xinjiang—China's Futuristic Anti-satellite Weapons, *The Trumpet*, April 05, 2019, https://www.thetrumpet.com/18883-laser-guns-in-xinjiang-chinas-futuristic-anti-satellite-weapons

44. PTI, China successfully launches second generation data relay satellite, *Economic Times*, April 01,2019, https://economictimes.indiatimes.com/news/international/world-news/china-successfully-launches-second-generation-data-relay-satellite/articleshow/68665863.cms?from=mdr
45. China launches world's largest single-dish radio telescope, domain-b.com, January 13, 2020, https://www.domain-b.com/industry/Aerospace/20200113_dish_radio.html
46. DIA Document, CHALLENGES TO SECURITY IN SPACE, January 2019, p. 20, https://www.dia.mil/Portals/27/Documents/News/Military%20Power%20 Publications/Space_Threat_V14_020119_sm.pdf
47. Dan Leone, NASA Spending Panel Chairman Keeps Focus on China, sapcenews.com, February 25, 2015, https://spacenews.com/nasa-spending-panel-chairman-keeps-focus-on-china/
48. Charlie Campbell, From Satellites to the Moon and Mars, China Is Quickly Becoming a Space Superpower, Time.com, July 17, 2019, https://time.com/5623537/china-space/
49. COMPETING IN SPACE, US National Air and Space Intelligence Centre, Publication Date: December 2018, https://media.defense.gov/2019/Jan/16/2002080386/-1/-1/1/190115-F-NV711-0002.PDF

7

Army and Maritime Aviation

"Air control can be established by superiority in numbers, by better employment, by better equipment, or by a combination of these factors."

—Carl Andrew Spaatz

"To have command of the air means to be able to cut an enemy's army and navy off from their bases of operation and nullify their chances of winning the war."

—Giulio Douhet

The PLA cutting its size by 50 per cent was expected and has implications. It is part of the structural transformation announced by President Xi Jinping and consistent with the military reforms introduced in 2015. The composition and disposition of Chinese military power is changing significantly, as China moves away from the 'strength in numbers' type of World War II composition. Known for large unproductive sloth and unmanageable lazy numbers, the PLA will soon shape into a more streamlined modern fighting force free from its internal institutional bureaucracy. Chinese military power has always believed in defeating the adversary without actually engaging in active combat. "To fight and conquer in all our battles is not supreme excellence; supreme excellence consists in breaking the enemy's resistance without fighting", said Sun Tzu,[1] the Chinese general and military strategist. If the same were not to happen, China would prefer subduing the opponent with overwhelming force at low cost. Attrition warfare was the basis of the PLA's war fighting principles.[2] Cutting the manpower-related flab, integrating theatre commands, instituting jointmanship in the form of combined arms warfare, transformed the PLA into a technologically advanced fighting force and these were thus the key aspects of military reforms.

The state military system upholds the principle of the Communist Party of China (CPC)'s absolute leadership over the armed forces. The party and the State jointly established the CMC that carries out the task of supreme military leadership over the armed forces. The CMC is the commanding organ of the PLA. To ensure the absolute leadership of the Communist Party over the armed forces, every level of party committee in the military forces implements the principles of democratic centralism.[3] In addition, division-level and higher units establish political commissars and political organisations, ensuring that the branch organisations are in line. This is seen as the key guarantee to the absolute leadership of the party over the military. The PLA reports not to the State Council but rather to two Central Military Commissions, one belonging to the state and one belonging to the party. In practice, the two central military commissions usually do not contradict each other because their membership is usually identical.

Forces under each military branch or force, academies and schools, scientific research and engineering institutions and logistical support organisations are also under the leadership of the CMC. Under the civilian chairperson and vice-chairmen of the Central Military Commission of the Communist Party of China and the Central Military Commission of the People's Republic of China, the members are all military generals. The PLA has a joint staff directly attached to the Central Military Commission (CMC), the highest leadership organization in the military. They were divided into 25 functional departments within the Central Military Commission.

PLA Military Structures

The PLA encompasses five main service branches: the Ground Force, the Navy, the Air Force, the Rocket Force, and the Strategic Support Force. Following reforms, the strength is 2.0 million personnel. The reductions are mainly from non-combat ground forces, which allow more funds to be diverted to naval, air, and strategic missile forces. This shows China's shift from ground force prioritization to emphasizing air and naval power with high-tech equipment for offensive roles. The PLA has closely watched closely the performance of US forces in Afghanistan and Iraq. They have learnt from the US military about joint operations, network-centric operations, asymmetric warfare, C4ISR, and hi-tech weaponry.

Meanwhile, PLAN Headquarters controls the North, East, and South Sea fleets. PLAAF Headquarters exercises control through the air commanders of the five theatre commands. Nuclear forces are directly subordinate to the Joint Staff Department through the Strategic Forces commander and political commissar. The Joint Staff Department in Beijing could assume direct operational control of any main-force unit at will. Thus, the Joint Staff Department exercises operational control of the main forces, and the military region commanders control the regional forces. Among the deputy commanders of a major Theatre Command, there is always one who is assigned to take charge of intelligence work, and the intelligence agencies under his charge are directly affiliated to the headquarters and the political department of the military region.

Military Strategic Guidelines

China increasingly seeks to leverage its growing economic, diplomatic, and military influence to establish regional pre-eminence and expand its international influence. In 2015, China's leadership directed the PLA to be capable of fighting and winning "informatized local wars" with an elevated emphasis on "maritime military struggle," adjusting its guidance on the type of war the PLA should be prepared to fight.[4] China promulgated this revision through its "military strategic guidelines," the top-level directives derived from China's military strategy that prescribe concepts, assess threats, and set priorities for planning, force posture, and modernization. China expects significant elements of a modern conflict to occur at sea. China's leadership has adjusted its national military strategic guidelines about how to fight local wars.

People's Liberation Army Rocket Force (PLARF)

The 100,000 strong People's Liberation Army Rocket Force (PLARF) is the main strategic missile force of the PLA. It controls China's nuclear and conventional strategic missiles. China's total nuclear arsenal size is estimated to be between 100 and 400 (thermo) nuclear weapons. The PLARF has approximately six ballistic missile divisions (missile corps bases). They are independently deployed in different theatre commands and have a total of 15 to 20 missile brigades. The PLARF has solidified China's missile forces as a critical element of China's evolving strategic deterrent posture, portends continued significant investment in PLARF modernisation, and points to a more concerted effort to integrate PLARF capabilities into more effective PLA-wide joint operations—all key developments in the fundamental reshaping of China's approach to strategic deterrence.[5] The PLARF has two key missions: strategic deterrence and war-fighting. The PLARF is the "core force of China for strategic deterrence" with the responsibility for "deterring other countries from using nuclear weapons against China." At the same time, the PLA Navy and PLA Air Force are seeking a larger nuclear role through the development of sea and air platforms for the delivery of nuclear weapons.[6]

As China invests more heavily in its space-related infrastructure to enhance its strategic capabilities, especially in support of PLARF operations, in defence of kinetic and non-kinetic attacks on China's land and space based assets, the PLARF, is to develop the capacity to defy, deter and/or, if necessary, defeat the USA in a relatively short, localised conflict. The conventional forces of the PLARF will likely be employed offensively at the very outset of a campaign. Precision strike missile systems such as the DF-21 and DF-26 would be used.

People's Liberation Army Strategic Support Force (PLASSF)

Founded in December 2015, the People's Liberation Army Strategic Support Force (PLASSF) is the newest branch of the PLA. It has a strength of 175,000, and is a combination of all support forces, and includes high-tech operational forces such as space, cyberspace, and electronic warfare units, independent of other branches of the military. China realised the need that integration of information technology could grant military superiority. It also realised that the USA's use of these technologies created dependencies, which could be exploited during wartime. Practically, for China, cyber and information technology were the only domains to counter the asymmetry against the USA using critical nodes.[7] The Chinese therefore spent the late 1990s and early 2000s in enhancing their space and cyber capabilities.

The Science of Military Strategy 2013, which is the most important PLA textbook in recent times, demands China's strategic expansion in space and information domains. The literature also calls for merging of cyber, space, and electromagnetic capabilities. The SSF's creation paved the way for integration in these three domains. The reforms also brought the SSF under the CMC's direct command, giving President Xi Jinping complete control over it.[8] The SSF is broadly divided into two parts, the Space Systems Department (SSD) and the Network Systems Department (NSD). While the SSD is responsible for missions in space, the NSD looks after cyber and electromagnetic warfare. The SSF's mandate is to collect and harness information from outer space, cyber domain, and electromagnetic system and support for the forces for battlefield environment, information, communications, information security, and new technology testing. The SSF's information and technology support allows the PLA to enhance its strike capabilities.

PLA Special Operations Forces (PLASOF)

China's special ground force is called PLASOF. China's counterterrorist unit is drawn from the police force rather than the military. The Chinese Army fields a large number of special operations groups. China has developed a force capable of carrying out long-range air-borne operations, long-range reconnaissance, and amphibious operations. The size of the Special Operations Forces is estimated to be between 7,000 and 14,000 troops. The force's intended combat role is as rapid-response units in the event of a limited regional war under high-tech conditions. They also carry out commando, counter-terrorism, and intelligence gathering operations. The force supposedly receives army, air force, and naval training, including flight training, and is equipped with "hundreds of high-tech devices." The force's officers have completed military staff colleges, and 60 per cent are said to have university degrees. Chinese media claims, "over 4,000 soldiers of the force are all-weather and versatile fighters and parachutists who can fly airplanes and drive auto vehicles and motor boats". The PLAN South Sea Fleet and two Marine brigades have subordinate SOF units, as does the PLAAF 15th Airborne Army and Second Artillery Force (SAF).[9]

Theatre Commands

In 2016, China's territory was divided into five theatre commands instead of the earlier seven military regions. This reflected a change in their concept of operations from primarily ground-oriented to mobile and coordinated movement of all services. The five new theatre commands were Eastern, Western, Northern, Southern, and Central. The military reforms also introduced a major change in the areas of responsibilities. Rather than separately commanding their own troops, service branches are now primarily responsible for administrative tasks (like equipping and maintaining the troops). It is the theatre commands now that have the command authority. This should, in theory, facilitate the implementation of joint operations across all service branches.

Importance of Element Air

High economic growth did help Beijing invest a great deal in aerospace and maritime power. China also realizes that maritime and air powers enable it to project military strength beyond China's shores and land boundaries. This is very critical to the nation's first priority of re-uniting 'Taiwan', the Republic of China (ROC). It also helped transform the thinking in the PLA towards modern warfare. The infantry-centric PLA has also realised the importance of modern technology to win wars. Modern sensors, weapons, and use of air power are attributes PLA and PLAN are looking for investing. De facto, the PLAAF continues to be a part of the PLA as the name itself suggests but with the PLA having its own aircraft outside of PLAAF in a way makes their air element a separate entity. The PLAAF has also been wanting to rename itself as the 'Chinese Air Force', albeit without success yet. Meanwhile armies and navies the world over have realised the importance of Air Power, and are spending a significant part of their budgets to buy air assets vis-a-vis their organic tanks, guns, ships and submarines.

PLA GROUND FORCES AIR ELEMENT

The Big Picture

Beginning in the 1980s, the PLA tried to transform itself from a land-centred power to a more mobile, high-tech one capable of mounting operations beyond its borders.[10] The motivation for this were the new threats to China including the threat of declaration of independence by Taiwan, possibly with assistance from the USA. In 1985, the PLA reoriented itself to modernization, improving its fighting ability, and become a world-class force. The PLA has focused on increasing mechanization and informatization so as to be able to fight a high-intensity war. Chinese military planners call for short decisive campaigns, limited in both their geographic scope and their political goals. More attention is given to reconnaissance, mobility, and deep reach. This new vision has shifted resources towards the PLAN and PLAAF. The PLA is also actively preparing for space and cyber-warfare.

Therefore, China began acquiring advanced weapons systems from Russia, including the Su-27 and Su-30MKK and latest submarines, and new classes of destroyers and frigates. The PLAAF designed its very own J-10 fighter aircraft and a new stealth fighter, the J-20. The PLAN launched the new Jin Class nuclear submarines capable of launching nuclear warheads that could strike targets across the Pacific Ocean. It also has two aircraft carriers, one commissioned in 2012 and a second launched in 2017. According to the Stockholm International Peace Research Institute (SIPRI), China became the world's third largest exporter of major arms in 2010–14, an increase of 143 per cent from the previous five years.[11] China's share of global arms exports still remains around 5 per cent. A nearly 70 per cent of Chinese exports went to Pakistan, Bangladesh and Myanmar. Military spending in the People's Liberation Army has grown about 10 per cent annually over the last 15 years. The military budget announced in March 2019 was US$ 177.6 billion.

Another area is China's rapidly growing need for amphibious operations. "This expansion and upgrade has been part of the broader military overhaul taking place in recent years", said Li Jie, a naval expert based in Beijing.[12] There are more and more maritime interests to defend, especially islands and reefs. However, it is more important to have a stronger amphibious combat force to initially dominate the South China Sea, and later force projection outwards into the Pacific and the Asian littoral states. Expanding and upgrading the marine corps means better equipment, more specialised exercises, and systematic collaboration with other parts of the navy. Song Zhongping, a military commentator in Hong Kong, said that as China sought to build up a blue-water navy that can operate globally, new missions would emerge—such as security, counterterrorism and evacuation of Chinese nationals overseas—and they would fall to the marine corps.[13] Taking a leaf from the US Marine Corps, China would one day need expeditionary force capability. China's marine corps was first established in 1953 within the PLAN and its divisions now include amphibious infantry, armoured troops, artillery troops, air defence, engineering, scouts and special combat troops.

Air Elements PLA Ground Force

The PLA Ground Force operates mainly the attack and utility helicopters and small transport aircraft for communication duties. They have many small UAVs. The rough holdings as per Military Balance 2020 are in Table 1.

Table 1

Type	Origin	Class	Role	Numbers
Z-8	China	Rotorcraft	Transport	53
Z-9	China	Rotorcraft	Transport	255
WZ-10	China	Rotorcraft	Attack	91
Z-11	China	Rotorcraft	Attack/Patrol	46
Z-19	China	Rotorcraft	Attack/Patrol	105
Mi-8/17	USSR	Rotorcraft	Transport	250
Mi-26	Russia	Rotorcraft	Heavy Lift	4
Eurocopter AS 350	France	Rotorcraft	Training	53
Eurocopter EC120	France	Rotorcraft	Training	93
Aerospatiale Gazelle	France	Rotorcraft	Utility	8
Sikorsky S-70	USA	Rotorcraft	Transport	20
Y-7	China	Propeller Fixed Wing	Transport	7
Y-8	China	Propeller Fixed Wing	Transport	4
Y-9	China	Propeller Fixed Wing	Transport	2

WZ-10 In Western Theatre Command[14]

China's military has equipped all its ground forces with advanced WZ-10 combat helicopters which will be used to target battle tanks and air-to-air combat missions, a strategic move that could have implications for India. Several WZ-10s have been delivered to an aviation brigade of the PLA's 13th Group Army under the WTC. This means that all the Army's aviation units now have this advanced attack helicopter. Senior Colonel Xu Guolin, deputy chief of the PLA Army's Aviation Equipment Bureau, said that all the group armies will have at least one aviation brigade or regiment. Wu Peixin, an aviation analyst in Beijing, said the PLA Army now has a strong force of dedicated combat helicopters, thanks to the service of the WZ-10 and WZ-19, another attack helicopter that is less powerful than the WZ-10. "The Army now needs more medium-lift, multipurpose helicopters such as the US Army's Sikorsky UH-60 Black Hawk," he said. "This helicopter is capable of performing both combat operations and transport tasks." Gao Zhuo, a military observer in Shanghai, said the PLA Army needs at least 3,000 helicopters, especially heavy-lift transport types and multipurpose models.

WZ-10 Capabilities

The WZ-10 is a two-crew attack helicopter with a maximum take-off weight of 7,000 kg, and is powered by two Chinese WoZhou-9 engines of 1,000 kW (1,300 hp) each. It has been co-designed with Kamov Helicopter Division of Russia. It has a maximum speed of 270 km/h, and a range of 800 km. Its service ceiling is 6,400 m. It has one 23 mm gun, and on its four external hard-points, can carry 1,500 kg useful load, which could be 57 mm or 90 mm unguided rocket pods or up to 16 HJ-10 air-to-surface anti-tank or anti-helicopter missiles. It could carry up to four PL-5, PL-7, or PL-9 air-to-air missiles. The avionics include a Chinese YH millimetre-wave fire-control radar, Helmet mounted sight with Night Vision Goggles, the BM/KG300G self protection jamming pod, Blue Sky Navigation pod, KZ-900 reconnaissance pod, and YH-96 electronic warfare suit. China has been claiming the helicopter to be comparable to the Russian Mi-28 and American Apache 64.

Meanwhile, India has inducted the Boeing AH-64E Apache in September 2019. One unit of these is operational in the North and one in North East sector. The Indian Army has also ordered the same helicopter. This variant evolved in 2012, with increased capabilities, and features improved digital connectivity, the Joint Tactical Information Distribution System, more powerful T700-GE-701D engines with upgraded transmission. It has the capability to control UAVs, has full IFR capability, and improved landing gear. The new composite rotor blades increase cruise speed, climb rate, and payload capacity. The updated Longbow radar has an over sea capacity, potentially enabling naval strikes and an AESA radar. AH-64Es have a data-link communicating on C, D, L, and Ku frequency bands to transmit and receive data and videos, including from UAVs.

WZ-10 Helicopter.
Picture Source: defencetalk.com

In comparison to the WZ-10, the Apache 64E has a maximum take-off weight of 10,432 kg.[15] The two engines have 1,409 kW (1,890 shp) each. The maximum speed is 293 km/h, and the range is 476 km with Longbow radar mast. The service ceiling 6,100 m. The armament includes a 30 mm gun, with 1,200 rounds as part of the Area Weapon sub-system. It has four external hard-points, and a station on each wingtip for an AIM-92 Stinger twin missile pack. The weapons include Hydra 70 mm air-to-ground rockets, AGM-114 Hellfire air-to-ground missiles, AIM-92 Stinger air-to-air missiles. Avionics include the Longbow fire-control radar, and the Integrated Helmet and Display Sighting System where either the pilot or gunner can slave the helicopter's 30 mm gun to their helmet. The gun can also be controlled via the Target Acquisition and Designations System. A thermo-graphic camera can precisely locate muzzle flashes. Clearly, the Apache is a larger and more operationally powerful helicopter for operations in Ladakh or the Sikkim Plateau.

PEOPLE'S LIBERATION ARMY NAVY (PLAN)

Until the early 1990s, the People's Liberation Army Navy (PLAN) performed a subordinate role to the PLA Land Force. However, since then it has undergone rapid modernization. The 240,000-strong force is organized into three major fleets; the North Sea Fleet with headquarters at Qingdao on Yellow Sea, overlooking Korean peninsula; the East Sea Fleet headquartered at Ningbo, south of Shanghai on East China Sea, overlooking Japan; and the South East Fleet headquartered in Zhanjiang, east of Hong Kong, overlooking the South China Sea. Each fleet consists of a number of surface, sub-surface vessels, coastal defence and marine units and naval air assets. The navy includes a 15,000 strong Marine Corps (two brigades), and a 26,000 strong Naval Aviation Force operating several attack helicopters and fixed-wing aircraft. As part of the modernization, the PLAN is in the stage of developing a blue water navy.

PLAN's 70th anniversary fleet review and celebrations were held on 23 April 2019. A 32-strong Chinese fleet led by Asia's largest destroyer, and PLAN's new addition, the Type 055, were joined by 18 foreign warships from 13 countries.[16] The naval review began with two Type 094 nuclear-powered submarines which can carry 16 nuclear-armed missiles. It is believed to be the second most powerful vessel of its kind after the US Navy's Zumwalt-class destroyer. The Liaoning, China's only active aircraft carrier, and its strike group closed the event—and the J-15 fighter jets on its deck were the only planes that visible during the entire parade. A total of 39 aircraft, including J-10 fighter jets, helicopters and strategic bombers, were due to take part in the event but poor visibility meant that none of them could be seen from the decks of the participating vessels.

China now has more warships than the United States Navy. The People's Liberation Army Navy (PLAN), informally known as the Chinese Navy, recently hit a landmark number of 300 ships—thirteen more than the U.S. Navy.[17] The U.S. Navy trails at 287 hulls, Russia has 83 hulls, the U.K. 75 hulls, and Australia at 48 hulls. According to the CSIS, the Chinese Navy has more ships than "Germany, India, Spain, and the United Kingdom" together.

Chinese Aircraft Carrier Program

A key element for PLAN's force projection and control in the Pacific and Indian oceans and the drive to catch up with the US Navy is to build four nuclear aircraft carriers. Beijing expected to have at least six aircraft carrier battle groups by 2035 after it prioritized modernizing its navy.[18] Nuclear-powered carriers are planned to be equipped with electromagnetic launch catapults similar to those of the USA. Four of at least six of the aircraft battle groups China plans to have in the water by 2035, will be nuclear-powered, according to Chinese naval expert Wang Yunfei, a retired PLAN officer. The hardware is expected to closely match US aircraft carrier

Liaoning Aircraft Carrier.
Picture Source: lowyinstitute.org (Photo VCG via Getty)

technology but PLAN would still lag in real combat experience. It is a long way to go since China has only one aircraft carrier in service, the Liaoning, commissioned in 2012, and the Type 002, the first carrier built in the country, which is still being tested.

Construction of its next conventional diesel-powered aircraft carrier, the Type 003, the first equipped with electromagnetic launchers, began in 2017. President Xi Jinping has ordered the PLA to modernize by 2035 and become a top-ranking fighting force by 2050.[19] China's fleet of aircraft carriers would expand to reflect its global standing. The budget for the carrier projects is not being cut despite economic slowdown and a trade war with the USA as the PLA plans to have four nuclear-powered carrier battle groups on the front line by 2035.

Chinese Aircraft Carrier Development History

The Chinese aircraft carrier development, production, and operation of program are primarily driven by the PLAN. China has ambitions to operate aircraft carriers since the 1970s, but at that time had never owned or operated an aircraft carrier. Since 1985, China acquired four retired aircraft carriers for study—the Australian HMAS *Melbourne* and the ex-Soviet carriers *Minsk, Kiev* and *Varyag*.[20] From the HMAS *Melbourne* China, learnt about the steam catapult, arresting equipment and mirror landing system by reverse-engineering a land-based replica of the steam catapult and landing system for training. The experience gained was later applied to the development of the J-15. The carrier was dismantled only in 2002. After detailed study, the *Minsk* and *Kiev*, though contractually meant to be scrapped, were converted into tourist attractions.

The 67,500 ton ex-Soviet 'Varyag' (Kuznetsov class) was only 68 per cent completed and floating in Ukraine when purchased through a private Macau tourist venture in 1998 and towed to Dalian Shipyard. The carrier had been stripped of all military equipment as well as her propulsion systems prior to sale. The entire refit exercise was kept under wraps. In mid-2007, China purchased a total of four sets of aircraft carrier landing systems from Russia, as confirmed by Russian manufacturers. In August, 2008, Mr. Huang Qiang, the speaker of the Commission of Science, Technology and Industry for National Defence, made a public announcement that China had mastered all the technologies for an aircraft carrier, and would build its own in the future when the time was deemed right.[21] Research and development on the planned carriers was being carried out at a military research facility in Wuhan. The actual carriers were constructed at Jiangnan Shipyard in Shanghai.

In August 2011, it was announced that the refurbishment of the *Varyag* was complete. In December 2011, an American satellite imaging company discovered the retrofitted *Varyag* undergoing trials in the Yellow Sea.[22] The carrier was named Liaoning after the Liaoning province of China and commissioned in September 2012 and handed over to the PLAN. In November 2012, the first landing on the carrier was successfully conducted with J-15. China confirmed that it was constructing a second carrier that would be built entirely with indigenous Chinese designs, similar to the *Liaoning*, and will also use a ski jump for take-off. *Liaoning* became combat ready four years later, in November 2016, and was given the Chinese designation of Type 001. In August 2018, the *Liaoning* underwent its first refit since its commissioning in 2012. The radar above the bridge and the air traffic control at the rear of the island superstructure went through changes.[23] It completed maintenance and modifications in January 2019.

In November 2013, China began fabricating the second aircraft carrier, the Type 001A 'Shandong', which was to be the first indigenous carrier. It was finally launched on 26 April 2017 and was called Type 002 at the time of commissioning. It was constructed at the Dalian Shipbuilding Industry. The Chinese government did not publicly confirm the ship's existence until construction was well underway. Satellite imagery showed the aircraft carrier in the early stages of hull assembly by March 2015. In December 2015, a Chinese defence ministry spokesperson confirmed that the ship was an aircraft carrier, stating that the design and construction work was under way.

The 002 is a conventionally powered ski jump carrier with a displacement of around 65,000 tons. It is known to be powered by oil-fired boilers and steam turbines, though previously there has been speculation that it could be gas-turbine propelled. With a bigger hangar, the new aircraft carrier can accommodate up to 32 J-15 fighter jets, compared to 26 on the Liaoning. Other redesigned elements include a bridge on two levels rather than one, with bigger windows. The Type 002 left her home port of Dalian for her third sea trial on October 28, 2018. It has been seen to have J-15 fighter jets and Z-18 helicopters on its deck.[24] Beijing-based naval expert Li Jie said that aircraft take-off and landing exercises would be conducted on the high seas after the Type 002 formally entered service. A waist catapult has been fitted to facilitate airborne early warning aircraft operation and air control. To compensate for the lack of a large AEW plane, the Chinese navy acquired Z-18J and Ka-31 helicopters that carry air-surveillance radars. Shandong was commissioned into PLAN service on December 17, 2019.[25] The "helicopters provide the PLAN a service-able sea-based AEW capability to help fill that critical gap until newer catapult-equipped aircraft carriers capable of operating fixed-wing AEW aircraft enter service," the US Defence Intelligence Agency (DIA) stated.[26]

The third aircraft carrier, known as Type 003, is to be an entirely different design from Types 001 and 002. It will have a displacement of around 85,000 tonnes.[27] Both an EMALS and a steam-powered catapult were constructed at the Huangdicun naval base for testing,[28] and it is believed to indicate that the Type 003 class as well as future PLAN carriers could possibly be CATOBAR carriers. The construction of the first Type 003 class aircraft carrier started in February 2017. It is claimed that two ships of the class are being built in Dalian Shipyard and Jiangnan Shipyard simultaneously. The Type 004 is planned to be even larger than the Type 003, and will also feature nuclear propulsion, and could have weapons like lasers and rail-guns. It is claimed that construction started in December 2017 at Jiangnan Shipyard. China plans to operate five aircraft carriers in the near future, including two nuclear-powered ones that would be launched around 2025 to meet the evolving strategic shifts, according to Chinese military experts.[29]

Currently planned Chinese carriers can be summarized in Table 2.

Table 2: Chinese Aircraft Carriers

Class	*Name*	*Builder*	*Launched*	*Commissioned*	*Status*
Type 001	Liaoning	Dalian Shipyard (Refit)	Acquired 1998	September 2012	Active
Type 002	Shandong	Dalian Shipyard	26 April 2017	Estimated End 2019	Ninth sea trial: from 14 November 2019
Type 003	TBD	Dalian Shipyard	Estimated 2020	Estimated 2023	Under Construction
Type 004	TBD	Jiangnan Shipyard	Estimated 2020	Late 2020	Under Construction since 2017

Carrier-Based Fighters

China initially intended to acquire Russian Sukhoi Su-33 carrier-based aircraft for its aircraft carriers. However, China later developed the Shenyang J-15, which was essentially a derivative of the Su-33, featuring Chinese technology and avionics from the J-11B program, and is currently in carrier service. On 25 November 2012, it was announced that at least two J-15s had successfully landed on *Liaoning*. The Flying Shark is claimed to be marginally superior compared to the F/A-18E/F Super Hornets flown by the U.S. Navy. Actually the J-15 still has issues related to its engines. The J-15 can carry only a limited payload (less than 12 tons) when taking off from the ski jump curved-deck of China's first two carriers.[30] There are unconfirmed reports of a J-15A catapult compatible fighter being tested on an Electromagnetic (EMALS type) catapult. The catapults and larger flight decks on China's next four carriers should address this problem. The PLA Navy is apparently interested in a step change to operating a stealth fighter. Meanwhile, the PLA Naval Air Force is planning to outfit their carriers with stealth fighters. China has two twin-engine land-based stealth jets for possible adaptation for carrier operations. The heavier J-20 that is already in service with the PLAAF or the smaller Shenyang FC-31 Gyrehawk. The latter bears a strong resemblance to the American F-35, a design reportedly stolen from BAe computers in 2007.[31]

Shenyang FC-31. *Picture Source: Wikipedia*
Location: China International Aviation & Aerospace Exhibition

Currently the FC-31 is officially for export, and would be re-designated as the J-31 if it ever enters Chinese service. The J-31 weighs 17 tons empty and 28 tons at maximum take-off weight, whereas a J-20 is 19 tons empty and max take-off at nearly 37 tons. In addition, the FC-31 is smaller with

17 metre-length compared to the 21-metre length of the J-20. Since the current J-15s are already larger than both at 22 metres are, so length may not be an issue. By mid-2019, analysts and media were increasingly convinced that the J-31 was more likely to be China's carrier-based stealth jet.[32]

A report in August 2019 indicated that defence insiders had indicated that China's Central Military commissions were secretly tilting towards the larger J-20 over the J-31.[33] Perhaps China is so inclined because the USA has been successfully using the larger F-14 Tomcat on its carriers. Despite the costs and larger size, the J-20's range and larger internal weapon bays gives it an advantage as a better interceptor and long-range strike aircraft. The J-20s claimed combat radius of 1,770 km on internal fuel is more than double the unrefueled range of the American Super Hornet and beats the F-35C by nearly 66 per cent.[34] This means a Chinese carrier could theoretically launch J-20s to strike U.S. carriers or bases while remaining out of range of retaliatory strikes. Another thing in the J-20's favour is that it's already in service and had matured from years of testing and operations, while only two FC-31 prototypes have been built. If China can spare enough funds, they may still want two carrier based stealth fighters. Only time will tell. Meanwhile the naval variant of J-20 would require folding wings, a tail hook for the arrestor cables, a strengthened landing gear, and some form of automatic carrier-landing systems. These modifications would add weight and decrease speed and range. Reports indicate that Chengdu is working on a shortened J-20 airframe to ensure compatibility with the catapult system on the forthcoming Type 003 carrier.[35]

Chinese media has also revealed a two-seat variant of the J-20 optimized for the strike and/or electronic warfare role, and possibly capable of carrier operations. Will this increase its radar cross-section is the question being asked, because no other stealth has a two-seat variant.[36] Both Chinese jets suffer from engine trouble. While China is struggling to develop the indigenous WS-15 turbofan engine for the J-20, it is managing with the Russian AL-31F turbofan, which can produce only 77 per cent of the required thrust. Meanwhile, the FC-31 continues to have the Russian RD-93 engine, while it also awaits completion of a WS-15, which is also intended for the JF-17 Thunder with Pakistan. Clearly, it will be at least a decade before China's naval stealth fighters are operational, so the J-15 will doubtlessly be the mainstay for some time.

Future Aircraft Carriers: USA and China

Both the USA and China have been working on future aircraft carriers to enhance their respective navies.[37] The USA has more than 70 years' experience developing aircraft carriers while China has been working on aircraft carriers for six years. Operating at full capacity, aircraft carriers can launch and recover up to 240 aircraft per day. The US's Gerald R. Ford class (Ford class), being built by Newport News Shipbuilding has been under construction since 2005. The first vessel was commissioned in July 2017 and is expected for deployment by 2022. Four more vessels are currently in various stages of completion, with commissioning dates set between 2024 and 2034. How do the future aircraft carriers compare on key metrics such as size, speed, armament and cost?[38] The USS *Gerald R. Ford* can carry more than 75 aircraft, while the future aircraft carriers, the USS *Enterprise* and USS *John F Kennedy,* can carry up to 90 combat aircraft. The Ford-class vessels are colossal, weighing 100,000 tons and have a length of 337m and a beam (flight deck) of 78m. China's Type 002 has the capacity to hold 38-40 aircraft in total, around half the capacity of the US's Ford class. Type 002 is slightly smaller at 315m with a beam (flight deck) of 71m, and is lighter at 70,000 tons with full load. While the Ford class can travel at speeds of 30 kts (55.6 km/h), China's Type 002 is slightly faster at 31 kts (57.4 km/h). The US's Ford class are powered by two nuclear reactors, which can create a total of 700 MW of electricity, a 25 per cent increase on its predecessor, the Nimitz class. China's Type 002 uses conventional steam turbines with diesel generators. China's Type 003 will be 85,000 tons, and Type 004 may be 100,000 tons.

The US Ford-class aircraft carriers can be upgraded with lasers and energy dissipating armour systems to protect against high-speed GPS-guided missiles. They will have dynamic armour should the smart weapons get past the lasers, says a RAND Institute report of 2005.[39] Ford class carriers will have enough power reserve to accept these defensive systems. China's future aircraft carriers also come with three HHQ-10 surface-to-air missile systems with 18 cells each. The HHQ-10 system can operate with full autonomy and has a reaction time of around 6-8 seconds.

China's Type 002 class program is estimated to cost $ 9 bn for two vessels. In contrast, the first ship, the USS *Gerald R. Ford* was estimated by Congress to cost $ 12.8 bn with an additional $ 4.7 bn in research and development costs. The second ship, the USS *John F Kennedy,* will be slightly cheaper, at $ 11.3 bn. In terms of cost then, the Chinese Type 002 vessels are much cheaper, at around one-third of the price of their US counterparts. The US Ford-class aircraft carriers will outmatch the Chinese Type 002 class vessels in all areas, except the Chinese ship being one knot faster. There are of course issues about Chinese aircraft carriers readiness times.

PLAN Makes Strides

PLAN has made tremendous strides over the past few years in transforming itself into a world-class force. Long-range operations and live-fire exercises have become regular parts of the Navy's training, which is focused on combat readiness and capability, Vice-Admiral Shen Jinlong, the Navy's commander in chief, and Vice-Admiral Qin Shengxiang, Navy political commissar, said in a joint article published in the latest *Qiushi Journal*, a flagship magazine of the Communist Party of China Central Committee.[40] The Navy's carrier battle group has gained initial fighting capability, and the transformation of the naval aviation force has been proceeding well, they said.

Ahead of Taiwan's presidential elections in January 2020, China conducted a major military exercise in the waters near Taiwan. Drills were held at both ends of the Taiwan strait. PLAAF and PLAN were involved in the exercise. The exercise was also in response to US approval of arms sales worth $ 2.2 billion to Taiwan in July 2019. Taiwan responded by conducting its own military exercises using F-16s armed with AGM-Harpoon missiles. This was followed by firing 117 medium and long range missiles into the sea. Clearly China has realised the importance of the air power possessed by PLAAF and PLAN.

Aircraft Carriers Implications: India

The rapid acquisition of aircraft carriers by China is of concern to India and Japan. Japan recently announced plans to convert 27,000-tonne Izumo-class helicopter destroyers into aircraft carriers, sparking concerns in China.[41] The Indian Navy, which is operating aircraft carriers since 1961, is currently operating only INS *Vikramaditya*. The indigenous 40,000 ton, INS *Vikrant* has had delays. In December 2019, it was reported that the engines onboard the ship were switched on. In February 2020, all the major structural and outfitting work was declared complete by the government. In July 2019, Vice Admiral Saxena, Controller of Warship Production and Acquisition in the Navy, said that the ship was expected to be delivered by 2021.[42] In response to a question in the Rajya Sabha (upper house of the

Indian Parliament), Sripad Naik, the minister of state for defence, stated: "Ship's targeted delivery was affected due to delay in supply of aviation equipment from Russia."[43] In March 2020 it was revealed that the Indian Navy after commissioning the aircraft carrier will deploy the ship at L&T's shipyard in Kattupalli near Chennai, because the planned naval base in Rambilli near Vishakapatnam on the East Coast was not ready yet.[44] The navy wants to lease a 260 m berth at Kattupalli shipyard, for eight years from 2022 to 2030 for interim berthing of the ship, by which time the naval base at Rambilli is expected to come up.

India's second indigenous aircraft carrier, the 65,000 ton INS *Vishal*, also called the Indigenous Aircraft Carrier 2 (IAC-2), capable of carrying 55 (35 fixed-wing and 20 rotary-wing) aircraft, is planned to be built by Cochin Shipyard Limited. An EMALS Catapult Assisted Take-Off But Arrested Recovery (CATOBAR) system is under consideration. In May 2015, the then Chief of the Naval Staff, Admiral Robin Dhowan, initially floated the possibility of nuclear propulsion, but this was later changed to an integrated electric propulsion system due to the complexities involved in developing a nuclear reactor with a capacity of 500 to 550 megawatts as that would possibly take 15 to 20 years. The Indian Navy also reached out to four international defence companies for suggestions with the design of *Vishal*. The carrier is expected to enter service in the early 2030s.

It can be seen that the Indian aircraft carrier program is way behind in terms of timelines, investments, and technology. China is already increasing its presence in the Indian Ocean region. A strong blue water Navy is required by India for it to maintain influence in the region. The Indian Navy plans to field three carrier battle groups under its Maritime Capability Perspective Plan. The 45,000 ton INS *Vikramaditya* is currently the only operational aircraft carrier at present. The warship is capable of carrying around 30 aircraft and helicopters and can move at a speed of over 30 knots. Moreover, having a capacity of more than 8,000 tons of diesel, INS *Vikramaditya* has an operational range of over 13,000 km.

The Indian Navy has been arguing that it should have one aircraft carrier each for the eastern and western sea boards with one aircraft carrier in reserve to take care of the time taken for refit and repair of any of the two carriers. With INS *Vishal* slated for 2035, the Indian Navy will at best have two carrier groups at any given time till then. There will be periods of refurbishments, refits, overhauls, and maintenance, reducing to a single carrier group. China which is focused on competing with the USA for global dominance and has nearly four times India's defence budget will keep pulling ahead in Naval Aviation assets and capability.

China's Port Acquisitions around India

Also of concern to India is that China has already operationalised its base at Djibouti in the Indian Ocean and has acquired Sri Lanka's Hambantota port on a debt swap. China has developed the Pakistan's deep water Gwadar port in the Arabian Sea. In late 2015, the port was officially leased to China for 43 years, until 2059. In November, China also clinched a deal to build a multi-billion dollar port at strategically important Kyaukpyu in Myanmar, in the Bay of Bengal. Chittagong in Bangladesh, Malacca in Malaysia, and Mombasa in Kenya are also ports where China has invested. The Indian Ocean region is a key recipient of China's investment in foreign commercial ports. Chinese scholars recognize the far-reaching strategic significance of these projects for the success of the maritime Silk Road. As of 2018, Chinese companies have participated in the construction and operation of 42 ports in 34 countries under the Silk Road scheme.[45]

According to Sun Degang, Associate Research Fellow at the Shanghai Institute of International Studies, China's port projects along the Indian Ocean will be a comprehensive test of China's economic capabilities, risk prevention and the ability to set the international agenda. The Chinese perspective on the Indian Ocean, according to Sun, is threefold: from the perspective of maritime security, the Indian Ocean is a fragile region; from the perspective of economic interests, the Indian Ocean is a key region; from a strategic perspective, the Indian Ocean enables China to implement an opening towards the West.[46] The challenge for China is to address the strong international concern caused by these activities in many countries, and especially in India. Sun Degang perceives India as a disruptive factor China must face. At the same time, China also faces varying degrees of interference created by the USA. Currently, Sri Lanka is about to sign an agreement with India and Japan to develop a deep-sea container terminal at the port of Colombo, next to a Chinese terminal. The notion that the deepening U.S.-India relationship poses a major challenge to China's port construction plans in the Indian Ocean is strongly shared by other Chinese scholars. They highlight how U.S.-India-China strategic competition and cooperation and in particular the "Indo-Pacific Strategy" have increased geopolitical risk on the construction of the Belt and Road Initiative (BRI).

China also wants to accelerate the construction of multi-level shipping supply bases along the Indian Ocean coast, for which Indian Ocean routes are important. It is not enough to rely solely on a Djibouti overseas security base, and there is a need to create supply bases in Pakistan, Sri Lanka, and Kenya.

Chinese scholars argue that even though New Delhi has reached a strategic agreement with the USA on maritime cooperation, it is possible to delay the progress of the cooperation between the two countries, which gives China abundant time to plan its own Indian Ocean policy.[47] China is currently focusing on establishing its influence in the Indian Ocean countries, and plans further participation in port projects in the region. China would try to optimise its bilateral relations with countries that receive Chinese investment for port development. China is also facing resistance from Balochis in Pakistan, and there is a threat from the Islamic State.

Options India

In view of the delays in aircraft carriers, India must look at other options for defence of its interests in the Indian Ocean Region (IOR). Considering the huge financial outlays required for carriers, it could be a good idea to create significant naval and aircraft assets in the island territories, especially the Andaman and Nicobar chain of islands. Besides, land-based aircraft with aerial refuelling would be able to cover distances up to 4,000 kilometres. Once positioned in these islands, they will be able to cover large areas of the Indian Ocean.

In recent years, India has recognized China's 'Malacca dilemma,' a reference to the long and insecure lines of communication through the northern Indian Ocean that China relies upon for much of its international trade—and its corresponding interest in improving the security of its Indian Ocean trade routes.[48] China, for its part, needs to reassure India on its port projects and the military aspects of its regional involvement, particularly in the South Asian littoral. Even as they recognize some of China's maritime security concerns, Indian policymakers must be prepared to counteract any attempts at militarization or conversion of Chinese infrastructure investments in South Asia into forward-basing facilities for the Chinese military. A failure to do so might impel other great powers to respond to China with their own military bases on the subcontinent's maritime periphery, a process that would pose adverse consequences for regional geopolitics and also erode India's influence in the neighbourhood.

REFERENCES

1. Brainy Quotes, https://www.brainyquote.com/quotes/sun_tzu_133291
2. Harsh V. Pant and Kartik Bommakanti, Military reforms have put China's combat capability way ahead of India's, Observer Research Foundation, February 08, 2019, https://www.orfonline.org/research/military-reforms-put-china-combat-capability-way-ahead-india-47909/
3. Dr. R.K. Sahay, *History of China's Military*, Chapter: Organisation of National Military Command, Vij Books India Pvt. Ltd., 2016.
4. Military and Security Developments Involving the People's Republic of China 2019, ANNUAL REPORT TO CONGRESS, p. 14, https://media.defense.gov/2019/May/02/2002127082/-1/-1/1/2019_CHINA_MILITARY_POWER_REPORT.pdf
5. Professor Bates Gill and Adam Ni, The People's Liberation Army Rocket Force: Reshaping China's Approach to Strategic Deterrence, Australian Institute of International Affairs, September 15, 2019 http://www.internationalaffairs.org.au/australianoutlook/the-peoples-liberation-army-rocket-force-reshaping-chinas-approach-to-strategic-deterrence/
6. Ibid.
7. Suyash Desai, PLA SSF: Why China will be ahead of everyone in future cyber, space or information warfare, *The Print*, December 31, 2019, https://theprint.in/opinion/pla-ssf-why-china-will-be-ahead-of-everyone-in-future-cyber-space-or-information-warfare/342772/
8. Ibid.
9. Kevin McCauley, PLA Special Operations: Forces, Command, Training and Future Direction, Publication: *China Brief*, Volume 15, Issue 16, Jamestown Foundation, August 18, 2015, https://jamestown.org/program/pla-special-operations-forces-command-training-and-future-direction/
10. Dr. R.K. Sahay, *History of China's Military*, Chapter: Organisation of National Military Command, Vij Books India Pvt. Ltd., 2016.
11. Reuters, China displaces Germany as world's third largest arms exporter, March 16, 2015, https://www.reuters.com/article/us-china-military-exports-idUSKBN0MC0QT20150316
12. Liu Zhen, China's navy expands marine corps into own unit 'to defend maritime interests', *South China Morning Post*, April 22, 2019, https://sg.news.yahoo.com/china-navy-expands-marine-corps-125506119.html
13. Ibid.
14. PTI, PLA equips all ground forces with new attack helicopters, *Economic Times*, July 12, 2018, https://economictimes.indiatimes.com/news/defence/pla-equips-all-ground-forces-with-new-attack-helicopters/articleshow/54199863.cms?from=mdr
15. Boeing Official Site, https://www.boeing.com/defense/ah-64-apache/
16. Minnie Chan, China puts a damper on navy's 70th anniversary celebrations as it tries to allay fears over rising strength, *South China Morning Post*, April 23, 2019, https://www.scmp.com/news/china/military/article/3007383/china-puts-damper-navys-70th-anniversary-celebrations-it-tries
17. Kyle Mizokami, China Now Has More Warships Than the U.S., *Popular Mechanics*, May 20, 2019, https://www.popularmechanics.com/military/navy-ships/a27532437/china-now-has-more-warships-than-the-us/
18. Minnie Chan and Guo Rui, China will build 4 nuclear aircraft carriers in drive to

catch US Navy, experts say. *South China Morning Post*, February 06 2019, https://www.scmp.com/news/china/military/article/2185081/china-will-build-4-nuclear-aircraft-carriers-drive-catch-us-navy

19. PTI, China to build 4 nuclear aircraft carriers to catch up with US Navy, experts say, *Economic Times*, February 06, 2019, https://economictimes.indiatimes.com/news/defence/china-to-build-4-nuclear-aircraft-carriers-to-catch-up-with-us-navy-experts-say/articleshow/67869301.cms?from=mdr
20. Dipanjan Roy Chaudhury, Virus outbreak impacts China's aircraft carrier plans, defence programs, *Economic Times*, Updated February 04, 2020, https://economictimes.indiatimes.com/news/defence/virus-outbreak-impacts-chinas-aircraft-carrier-plans-defence-programs/articleshow/73942231.cms?from=mdr
21. Chinese aircraft carrier programme, Wikipedia, https://en.wikipedia.org/wiki/Chinese_aircraft_carrier_programme
22. Associated Press, Chinese aircraft carrier photo captured by U.S. satellite company, Politico, December 14, 2011, https://www.politico.com/story/2011/12/chinese-aircraft-carrier-photo-captured-by-us-satellite-company-070442
23. Liu Zhen, China's Liaoning aircraft carrier gets brand new command centre, *South China Morning Post*, August 28, 2018, https://www.scmp.com/news/china/diplomacy-defence/article/2161582/chinas-liaoning-aircraft-carrier-gets-brand-new-command
24. David Donald, China Building a Carrier Force, AIN online, February 09, 2020, https://www.ainonline.com/aviation-news/defense/2020-02-09/china-building-carrier-force
25. Minnie Chan, First made-in-China aircraft carrier, the Shandong, officially enters service *South China Morning Post*, December 17, 2019, https://www.scmp.com/news/china/military/article/3042469/chinas-second-aircraft-shandong-carrier-officially-enters
26. David Axe, Is China's Second Aircraft Carrier A Floating Paper Tiger? *The National Interest*, December 18, 2019, https://nationalinterest.org/blog/buzz/chinas-second-aircraft-carrier-floating-paper-tiger-106121
27. David Donald, China Building a Carrier Force, AIN online, February 9, 2020, https://www.ainonline.com/aviation-news/defense/2020-02-09/china-building-carrier-force
28. Mike Yeo, China claims breakthrough in electromagnetic launch system for aircraft carrier, defensenews.com, November 9, 2017, https://www.defensenews.com/naval/2017/11/09/tech-breakthrough-chinas-next-carrier-could-feature-electromagnetic-launch-system/
29. PTI, China plans to have five aircraft carriers in future: Chinese experts, *Economic Times*, December 06, 2018, https://economictimes.indiatimes.com/news/defence/china-plans-to-have-five-aircraft-carriers-in-future-chinese-experts/articleshow/66966130.cms?from=mdr
30. Sebastian Roblin, Why China Is Bringing Stealth Fighters to Its Aircraft Carriers, *The National Interest*, March 18, 2020, https://nationalinterest.org/blog/buzz/why-china-bringing-stealth-fighters-its-aircraft-carriers-134212
31. Siobham Gorman, August Cole and Yochi Dreazen, Computer Spies Breach Fighter-Jet Project, *Wall Street Journal*, Updated April 21, 2009, https://www.wsj.com/articles/SB124027491029837401
32. Rick Joe, Beyond China's J-20 Stealth Fighter, *The Diplomat*, September 20, 2019, https:/

/thediplomat.com/2019/09/beyond-chinas-j-20-stealth-fighter/

33. Minnie Chan, China's navy 'set to pick J-20 stealth jets for its next generation carriers', *South China Morning Post*, August 27, 2019, https://www.scmp.com/news/china/military/article/3024584/chinas-navy-set-pick-j-20-stealth-jets-its-next-generation
34. Sebastian Roblin, Why China Is Bringing Stealth Fighters to Its Aircraft Carriers, *The National Interest*, March 18, 2020, https://nationalinterest.org/blog/buzz/why-china-bringing-stealth-fighters-its-aircraft-carriers-134212
35. Ibid.
36. Ibid.
37. Tala Husseini, Future aircraft carriers: Gerald R. Ford class vs Type 001A class, Naval Technology, May 22 2019, https://www.naval-technology.com/features/future-aircraft-carriers-us-china/
38. Ibid.
39. John Schank, Giles Smith, Brien Alkire, Mark V. Arena, John Birkler, James Chiesa, Edward Keating and Lara Schmidt, Modernizing the U.S. Aircraft Carrier Fleet, Rand Corporation report, p. 83, 2005, https://www.rand.org/content/dam/rand/pubs/monographs/2006/RAND_MG289.pdf
40. Zhao Lei, World-class PLA Navy is emerging, *China Daily*, April 23, 2019. http://www.chinadaily.com.cn/a/201904/23/WS5cbe6420a3104842260b7a8b.html
41. PTI, China plans to have five aircraft carriers in future: Chinese experts, *Economic Times*, December 06, 2018, https://economictimes.indiatimes.com/news/defence/china-plans-to-have-five-aircraft-carriers-in-future-chinese-experts/articleshow/66966130.cms
42. PTI, Indigenous Aircraft Carrier likely to go for basic trials next year, *Economic Times*, https://economictimes.indiatimes.com/news/defence/indigenous-aircraft-carrier-likely-to-go-for-basin-trials-next-year/articleshow/70138610.cms?from=mdr
43. Om Krishna, Delay in supply of equipment from Russia affected delivery of India's aircraft carrier, defencestar.in, February 12, 2020, https://www.defencestar.in/military-news/indian-navy-news/defence-news-india-updates-delay-in-supply-of-equipment-from-russia-affected-delivery-of-indias-aircraft-carrier-ins-vikrant/
44. ANI, Indian Navy to berth new aircraft carrier at private shipyard, *Livemint*, March 15, 2020, https://www.livemint.com/news/india/indian-navy-to-berth-new-aircraft-carrier-at-private-shipyard-11584281200203.html
45. Angela Stanzel, China Trends #2 – China's String of Ports in the Indian Ocean, Institut Montaigne, June 25, 2019, https://www.institutmontaigne.org/en/blog/china-trends-2-chinas-string-ports-indian-ocean
46. Ibid.
47. Ibid.
48. Zorawar Daulet Singh, Managing India-China relations in a Changing Neighbourhood. Centre for Policy Research, June 04, 2019, https://www.cprindia.org/policy-challenge/7848/foreign-policy-and-national-security

8

China Mentors Pak Military Aviation Industry

"China-Pakistan relationship is indeed special because it transcends the changes of the times and politics and represents a fine example of friendly state-to-state interactions."

—Li Keqiang, Chinese Prime Minister

"Not to have an adequate air force in the present state of the world is to compromise the foundations of national freedom and independence."

—Winston Churchill

A significant element of the Pakistan Air Force (PAF) riposte on 27 February 2019 after the Indian Air Force's (IAF) Balakot strike the previous day was the Sino-Pakistani JF-17 'Thunder' aircraft, the flagship joint defence project. "Pak-China friendship is higher than Himalayas, deeper than ocean, sweeter than honey, and stronger than steel"[1] is a statement attributed to many a Pakistani leaders. On January 26, 2015, Chinese Foreign Minister Wang Yi, during the conclusion of a two-day visit of former Army Chief Raheel Sharif to Beijing called Pakistan China's "irreplaceable, all-weather friend."[2] Maintaining close relations with China has been the central part of Pakistan foreign policy for decades. Pakistan was one of only two countries, besides Cuba, to offer crucial support to China after the Tiananmen Square protests of 1989. China and Pakistan also share close military relations, with China supplying a range of modern armaments to the Pakistani defence forces. China supports Pakistan's stance on Kashmir, while Pakistan supports China on issues related to Xinjiang, Tibet and Taiwan. Military cooperation has deepened, with joint defence production projects producing armaments ranging from fighter aircraft to guided missile frigates. Chinese cooperation with Pakistan has reached economic high points, with substantial Chinese investment in Pakistani infrastructural expansion including the Pakistani deep-water port at Gwadar, and the CPEC, which is the flagship of the BRI project.

"For Pakistan, China is the new America. From weapon systems to economic bailouts, to diplomatic support in various international fora, China is Pakistan's go-to country," writes Sushant Sareen.[3] The key complex dynamics in the relationship is the fact that for decades, the USA has been the principal patron of Pakistan and gave liberal financial aid and supplied significant high-end military hardware. Today China and the USA are competing for global dominance. The latter is wooing the more powerful India in the last two decades to counter-balance China. This has literally pushed Pakistan closer into China's arms.

One of the lynchpins of the Sino-Pakistan relationship has been defence cooperation. There are strong geopolitical and military ties between China and Pakistan. These ties are aimed primarily to counter Indian and American influence in the region. Interestingly, close U.S.-Pakistan relations for many years were also to repel Soviet influence in the area. In 1963, Pakistan ceded 5,180 square kilometres of strategic territory (Shaksgam valley) in northern Kashmir to close the border dispute with China. This marked the beginning of close ties between the two countries, including military cooperation. China helped the Pakistan army establish ammunition factories. Nuclear and missile technology support has been significant. Undoubtedly, the key element of defence cooperation has been cooperation in military aviation.

REGIONAL INTERNATIONAL LINKAGES

Vacillating Pakistan-USA Relations

Pakistan allied itself with the USA during the Cold war era against the Soviet Union and was an integral player in the Central Treaty Organization (CENTO) and Southeast Asia Treaty Organization (SEATO). The relations worsened during the socialist reign of the Pakistan People's Party (PPP) under Zulfikar Ali Bhutto. Yet Pakistan helped establish

US-China relations in the 1970s. Relations improved and deepened during 'Operation Cyclone,'[4] a US CIA program to arm and finance the Mujahideen in Afghanistan between 1979 and 1989, prior to and during the military intervention by the USSR. The program also helped strengthen militant Islamic groups that were favoured by Gen. Zia-ul-Haq in Pakistan. The Americans pushed in heavy funds to contain Russian influence or presence in the area. Relations once again soured after the collapse of the Soviet Union, when the USA approved sanctions against Pakistan by passing the Pressler Amendment[5] which was enacted against Pakistan for its nuclear weapons program in 1990. Pakistan once again assumed an important role in American geopolitical interests in the region following the September 11 attacks, and the subsequent 'War on Terror'. The USA named Pakistan a major non-NATO ally in 2002, which allowed the release of over $ 25 billion of aid to Pakistan.

Relations began to strain as both sides began to criticize each other's strategy in the 'War on Terror', with the United States government frequently accusing Pakistan of harbouring terrorists. The secret U.S. 'Operation Neptune Spear' in Abbottabad which resulted in the death of Osama bin Laden strained relations between the two countries. Public opinion in Pakistan frequently ranks the USA as one of its least favoured countries, and vice versa. Yet, Pakistan continues to occupy a strategic position in US interests in Central and South Asia. The USA is the second-largest supplier of military equipment to Pakistan after China, and is one of Pakistan's largest donors of foreign assistance. It has agreed to resume military aid to Pakistan as it needs Pakistan's support to end the war in Afghanistan. It is pertinent that the high-end military equipment with Pakistan continues to be of American origin, including the F-16 fighter aircraft.

Indo-Russian Bear Hug

Despite being non-aligned, during the Cold War, India and the Soviet Union had a strong strategic, military, economic and diplomatic relationship, which has continued with Russia as a special relationship. In the initial years, newly independent, but financial weak India preferred a socialistic approach. The Americans had literally pushed India into the Soviet fold by adopting and 'Us or Them' approach. Despite a few lean years after the Soviet Union broke up, the relations between Russia and India have always been privileged and strategic. A personal bonhomie shared between Prime Minister Modi and President Putin has again strengthened bilateral relations. The Indo-Russian strategic partnership is built on geopolitics, defence, nuclear energy, and space. Both countries have set a target of reaching $ 30 billion in bilateral trade by 2025, from about $ 10.9 billion in 2018, and January-September 2019 at $ 7.55 billion.[6] In addition to the UN, both countries are members of BRICS, G20 and SCO. Russia supports India for a permanent seat on the UNSC. India is the second largest market for the Russian defence industry. Approximately 68 per cent of Indian military hardware is of Russian origin, making it the biggest supplier. With the straining of relations between the USA and Russia after the annexation of Crimea, much to its dislike, Russia was pushed to close up with China. As India started leaning towards the USA in the last decade, and purchased major defence equipment, Russia also offered defence equipment to Pakistan albeit the contracts have been far and few.

Indo-US Dynamics

Indian Prime Minister Modi's very high profile 'Howdy Modi' sixth visit to the USA in September 2019, and President Trump's iconic visit to India in February 2020 further brought the two largest democracies closer. Convergence of global strategic interests and US support to India for membership in international groupings has been the lynchpin of the evolving relationship. Increase in bilateral trade and investment, co-operation on global-security matters, inclusion of India in decision-making on matters of global governance, upgraded representation in trade and investment forums (World Bank, IMF, APEC), admission into multilateral export control regimes (MTCR, Wassenaar, Australia Group) and support for admission in the Nuclear Suppliers Group (NSG) and joint-manufacturing through technology sharing arrangements have become key milestones and a measure of speed and advancement on the path to closer USA-India relations. In 2016, India and the USA signed the Logistics Exchange Memorandum of Agreement (LEMOA),[7] and later, the Communications Compatibility and Security Agreement (COMCASA).[8] Sale of state-of-the-art military hardware and regular joint military exercises have made the USA a strategic partner of India. Both are looking at the Indo-Pacific region with a similar strategic vision.

China's Neighbour Troubles

China, with the world's largest standing army and second largest defence budget, has traditionally been a hegemonic power and is still pushing the 23 boundary disputes with its 14 neighbours.[9] Territorial disputes with other countries include Taiwan; Yalu river enclaves with North Korea; Aksai Chin and Arunachal Pradesh with India; four pockets of land in Bhutan; the Japanese Senkaki Islands; Paracel Islands with Vietnam; Scarborough Shoal with Philippines; and a complex dispute over Spratly Islands with four countries.

String of Pearls

For over a decade China has been investing in the Indian Ocean region and using geopolitical influence through economic inducement, military supplies, airfields and port construction. Popularly known as the 'String of Pearls,'[10] India sees it as the encirclement of India. Chinese investments in Gwadar port in disputed Baluchistan in Pakistan, and the BRI road-rail corridor to connect through disputed Pakistan-occupied Kashmir has been questioned by India. Hambantota port in Sri Lanka, Chittagong and Sonadia in Bangladesh, a port in Sudan on the Red Sea, a port in Somalia and the first overseas military base in Djibouti are part of dominating the Indian Ocean region. China has for long been building roads and airfields in Myanmar and has developed a deep-water port and an electronic intelligence base at Coco Islands, just 18 km from India's Andaman Islands.

PAKISTAN-CHINA DYNAMICS

Pakistan-China Political and Economic Ties

Pakistan's geographical location makes it of great interest to many countries. The location allows influence in South Asia, West Asia, the Arab world, and Central Asia. China–Pakistan relations began in 1951 when Pakistan was among the first countries to end official diplomatic relations with the Republic of China (Taiwan) and recognised the PRC regime on mainland China. Since then, both countries have regularly exchanged high-level visits resulting in a variety of agreements. The PRC has provided economic, military, and technical assistance to Pakistan, and each country considers the other a close strategic ally. Boundary issues were resolved in 1963, military assistance began in 1966, a strategic alliance was formed in 1972, and economic co-operation began in 1979. After ceding the Trans-

Karakoram tracts to China in 1963, Pakistan began supporting China on Taiwan, Xinjiang, and Tibet, and also on human rights issues. Pakistan had served as China's main bridge to the Islamic world and also played an important role in bridging the communication gap between the PRC and the West by facilitating US President Nixon's historic visit to China in 1972. China has become Pakistan's largest supplier of arms and its third-largest trading partner. China supports Pakistan's civil nuclear power sector. China is today a central pillar of Pakistan's foreign policy. China and Pakistan also share close military relations, with China supplying a range of modern armaments to the Pakistani defence forces. China supports Pakistan's stance on Kashmir. The high point of Chinese cooperation with Pakistan is the substantial Chinese investment in Pakistani infrastructural expansion including the Pakistani deep-water port at Gwadar and the CPEC. While the CPEC will build modern infrastructure in Pakistan, the actual benefits will be accrued by China to get road-rail access to the Arabian Sea.

After serious boundary disputes emerged between China and India, and resulted in the Sino-India war of 1962, Pakistan was emboldened and made a failed attempt to wrest Kashmir from India in 1965. This failure pushed it even closer to China. In the 1971 Indo-Pakistan war, China did not come in support of Pakistan even when East Pakistan separated to become Bangladesh. Sino-Pakistan bonhomie has increased considerably since then. Even though China helps Pakistan on many fronts, including military and strategic hardware, China has not physically intervened on Pakistan's behalf against India during any conflict, including Kargil. China though continues to support Pakistan in the United Nations to prevent Pakistan to be declared a terrorist state.

Pakistan has also served as a conduit for China's influence in the Muslim world. China and Pakistan reached a comprehensive nuclear co-operation agreement in 1986. China considers its close relations with Pakistan as a counter-weight against the fast developing close relations between the USA and India. China also has consistently supported Pakistan on regional issues. This anti-India alliance has been used to benefit both nations on the diplomatic, economic and military fronts. China supported Pakistan in UN General Assembly (UNGA) in September 2019 on Kashmir. Pakistan has signed a 40-year lease with China allotting over 923 hectares of land for Beijing to develop a massive special economic zone near the deep-sea port of Gwadar. Goods from China would transit 2000 kilometres south from Xinjiang to the sea allowing China faster access to the Middle East, Africa and Europe. China hopes the CPEC will develop land-locked Xinjiang and make it easier to contain unrest. For Pakistan, China is a low-cost-high-value deterrent against India. China is today a major arms exporter and finds Pakistan a good route to showcase its products.

China-Pakistan Military Relationship

The current state of China-Pakistan military ties is best summed up by Catherine Wong.[11] China-Pakistan military are set to get even closer as they eye new alliances from jointly developed jets to high-level talks on regional security issues, Beijing and Islamabad are working side by side. The JF-17 Thunder is a low-cost combat aircraft designed in China and assembled in Pakistan. It is set to replace Pakistan's ageing fleet of Dassault Mirage III/5 fighter jets by 2020, Wong writes. There are reports that China is planning to build its second offshore naval base near a strategically important Pakistani port. Once dubbed 'iron brothers' by the now-retired vice-chairman of China's Central Military Commission, Fan Changlong, the two countries have steadily increased cooperation on military and defence matters in recent years as they seek to counter the perceived threat from rivals India and the USA. Pakistan has carried out a number of joint military exercises with China.

Beijing agreed to sell Pakistan eight Type 039B (upgraded) diesel-electric attack submarines by 2028 in a deal valued at between $ 4 billion and $ 5 billion. The deal was finalised on 23 July 2015. Four of the eight submarines are to be built in Pakistan. It includes transfer-of-technology to construct the vessels. The Chinese and Pakistani armed forces regularly conduct joint military exercises, intended to develop inter-operability between the forces. The two countries also hold exercises to counter terrorism. It is of great interest for China as its very restive Xinjiang region is close to Pakistan. China hopes that stability in Afghanistan and Pakistan will allow it to control Xinjiang better. It will also provide more security along the CPEC.

Analysing the trends in China-Pakistan arms transfer, Ambuj Sahu (ORF) wrote on 14 June 2019,[12] that almost 75 per cent of tanks being used by the Pakistan Army and 65 per cent of the aircraft used by the Pakistan Air Force (PAF) are of Chinese origin. China also allegedly had a vital role in technology transfer to Pakistan for its nuclear and missile programmes. In the last five years Pakistan's military hardware imports from China has grown to 69 per cent of the total. Airborne systems dominate Chinese hardware transfer. The political stability of Pakistan has been a primary concern for Chinese plans to succeed. Similarly, the USA-Pakistan arms trade has gradually declined, and is at an all-time low. It reflects the hardened stance of the Trump Administration on Pakistan due to its failure to crack down against terrorist outfits on their territory. Although Pakistan remains the largest arms market for Chinese hardware, China has been looking to find other markets. This coincides with the Chinese bid to compete with the USA for the global superpower spot. The heavy military and economic dependence on China forces Rawalpindi to comply with Beijing's terms, which sometimes make the Pakistani establishment uncomfortable.

China's official position remains, "Militarily, China eyes pragmatic and effective cooperation with Pakistan, which is in the front line of the fight against international terrorism. The military exchanges are not directed against any third party and contribute to peace and stability in both the region and the whole world", a Xinhua commentary said.[13]

Military and Defence Production Links

In the initial decades after the country's formation, Pakistan's military depended almost entirely on American armaments and aid, which increased considerably during the Soviet war in Afghanistan. When the Pressler Amendment suspended all American military assistance, many Pakistanis saw this as a betrayal. Consequently, since 1990, Pakistan has tilted towards China for military and economic cooperation. Pakistan found China as a more reliable ally over the long term. China helped Pakistan set up munitions factories and upgrading the existing ordnance factory at Wah near Rawalpindi. China allowed licensed production of the MBT-2000 (Al-Khalid) tank which was essentially a Chinese variant of the Russian T-90 over 450 of which have been produced. Al-Zarrar is another tank in licence production. China has also delivered many Type 59, Type 69 and Type 85 tanks. The armies of the two nations conduct joint exercises regularly. The Pakistan army has Chinese-designed short and medium-range ballistic missiles, including the Shaheen series with a range up to 2,500 km. China has also built a turnkey ballistic missile manufacturing facility near Rawalpindi and is constructing four submarines for the Pakistan Navy. All these projects involve transfer of technology. In 2008, a Chinese F-22P frigate joined the Pakistan Navy. In spite of serious global concerns, state-controlled Chinese military related technology, infrastructure and equipment continue to be the major portion of the $ 9 billion trade between the two countries. On April 19, 2015, China cleared the sale of eight conventional submarines to Pakistan worth $ 5 bn, in the biggest ever arms sale by China in its history.[14] China has supplied Pakistan with nuclear technology and assistance, including perhaps the blueprint for Pakistan's nuclear bomb. Chinese security agencies were reportedly in the know of transfer of nuclear technology by Pakistan to Iran, North Korea, and Libya. After India secured a nuclear deal with the USA, China agreed to set up two nuclear power stations in Pakistan.

THE MILITARY AVIATION LINKAGE

China's Military Aviation Production Support to Pakistan

Starting in 1965, China began supplying 253 F-6 aircraft, an air defence version of the Russian MiG-19. One squadron of Harbin H-5, a Chinese version of the Russian Ilyushin IL-28 was formed in the early 1970s. China helped establish 'Heavy Industries' at Taxila in 1971 for equipment rebuilding and in 1973, the Pakistan Aeronautical Complex at Kamra North of Islamabad. From 1983 onwards, the PAF received 55 A-5Cs ground attack variants of the MiG-19 and 186 F-7 aircraft, the Chinese equivalent of the MiG-21. In 2007, as part of a joint-project, China rolled-out a 'designed for Pakistan' fighter JF-17 'Thunder'. Currently, the PAF has over 100 of these and the strength is likely to increase to 300 some day. This program is also a signal to the USA that Pakistan has others to bank on for its security needs. The PAF had shown an interest to buy at least 36 Chengdu J-10 'Vigorous Dragon' fighters, but later changed its mind and would now go in for the JF-17 Block III.[15] The Block III would include AESA radar, helmet-mounted display, avionics improvements and possibly an internal infrared search and tracking (IRST) system, and perhaps some reworking of the airframe. A two-seat version will also be part of the Block III package.[16]

A large number of missiles and guided bombs have been supplied by China. Pakistan and China have also been working on the production of Unmanned Aerial Vehicles (UAV) including the Caihong-5 and Wing Loong-I. In October 2018, it was announced that China and Pakistan would jointly produce 48 Wing Loong II UAVs for use by the PAF.[17] Six ZDK-03 Chinese AWACS based on the Shaanxi Y-8F-600 cargo aircraft have been inducted. The ZDK-03 is a Pakistani variant specifically designed for the

Pakistan Air Force (PAF). The aircraft features an AESA radar with a range greater than the air force's Saab 2000 Erieye AEW&C radar.[18]

The fleet of Chinese Shenyang FT-5 and American T-37 trainers is being replaced by Chinese designed K-8 Karakorum intermediate jet trainers.[19] Sixty are currently in service and more are on order. The PAF has also received four CH-4 recce-cum-strike drones which can carry up to four PGMs and reportedly have an endurance of 30 hours. The Chinese PL-12 (SD-10) is the primary BVR air-to-air weapon. Short range, infra-red homing missiles include the Chinese PL-5E and PL-9C. The PAF is also seeking to arm the JF-17 with a fifth generation close-combat missile such as the PL10E IRIS-T or A-Darter.

Despite Chinese pledges to the contrary, it has continued to provide Pakistan with specialty steels, guidance systems and technical expertise in the latter's effort to develop long-range ballistic missiles. The M-11 is a copy of the DF-11 and the Hatf, Shaheen and Anza series of missiles have been built with Chinese assistance.

PAF K-8 Karakorum Intermediate Jet Trainer.
Picture Source: XairForces

JF-17 – The Success Story

The JF-17 Thunder or CAC FC-1 Xiaolong is a light-weight single-engine, multirole combat aircraft designed by China and produced jointly by the PAC and the Chengdu Aircraft Corporation (CAC) of China. The aircraft can be used for aerial reconnaissance, ground attack, and air interception. This fly-by-wire, fighter is powered by a Guizhou WS-13 or Kilmov RD-93 afterburning turbofan engine. It has a top speed of Mach

1.8. While the Chinese announced that flight testing of a JF-17 equipped with the indigenous WS-13 engine had begun, Pakistan decided to continue to use the RD-93 engine on its fighters.[20] China had contracted to buy 100 RD-93 engines from Russia for the FC-1 (JF-17), with an option to buy another 400.[21] The aircraft could later be powered by the Chinese indigenous Guizhou WS-13 engine. The aircraft has wide-angle Head Up Display, aerial refuelling, a data-link, and KLJ-7 Doppler radar. The aircraft has an electronic warfare suite. It can carry 6,700 lb (3,100 kg) external load on seven hard-points. The weapons are mostly Chinese, and include the Gsh 23 mm gun, PL-5 short-range air-to-air missile, LS-6 'Thunderstone' GPS-guided glide bombs, and YJ-12 supersonic and YJ-83 subsonic anti-shipping missiles. The PAF maintains one squadron in the maritime strike role. It had ordered 600 Chinese PL-12 radar-guided beyond-visual range (BVR) missiles with a range of around 80 km. The Chinese claim that the missile is comparable to the American AIM-120 AMRAAM and the Russian R-77.

Aircraft costs have been kept low by borrowing technologies developed for the Chinese J-10 fighter. The JF-17 is gradually becoming the backbone of the PAF. In 2015, the PAC produced 16 JF-17s. Currently Pakistan is believed to have the capacity to produce 25 JF-17s per year. Work share wise, 58 per cent of the airframe is Pakistani and 42 per cent Chinese. As of end 2019, Pakistan operates nearly 135 JF-17s in five operational squadrons, plus a testing and training unit. Nearly 70 jets are of Block 1 Type, and 62 are Block Type II. Aerial refuelling was introduced in Block II.[22] The last three JF-17 Block II aircraft were delivered to the PAF in June 2019. Meanwhile in May 2019, China delivered the first overhauled JF-17 back to the PAF.

The Block III Variant

A Block III variant of the JF-17 is under final development. It will have the Chinese KLJ-7A active electronically scanned array (AESA) radar, digital fly-by-wire flight control system, a new helmet mounted display, network-centric warfare capability, a holographic Head Up Display, an infra red search and track system, new electronic warfare systems, weapons upgrade and a radar cross-section reducing 'pseudo-stealthy' airframe. Pakistan is reportedly also still considering the Leonardo's Grifo-E AESA radar system. The aircraft is initially armed with two PL-5EII short-range air-to-air missiles (AAMs). The missile has a dual band, multi-element detector as well as a laser proximity fuse and an all-aspect capability. The PAF has ordered 900. It has the SD-10 (BVR),[23] C-802 anti-ship standoff, REK/IREK air-surface standoff with multiple warhead/seekers/roles and then all the GP bombs, CM-400 and a host of other A/G weapons for precision strikes. All that with the integrated ASEL Pod for targeting.

JF-17 'Thunder'.
Image Source: Wikipedia

The PAF claims it has tested the larger and longer range and more sophisticated PL-15 AAM don't seem to be true.[24] By incorporating a ramjet engine, the missile range could reach 150-200km.[25] The first two Block III aircraft have reportedly already entered production at the Aircraft Manufacturing Factory (AMF) final assembly line at PAC Kamra,[26] The Block III is being called a 4th generation-plus fighter by some. The JF-17 constitutes the 'backbone' of PAF and after the February 2019 engagement, claimed to be 'battle proven'. The AESA and PL-15 combination, if ever, with a 150-km range missile, PAF claims will outrange all IAF aircraft other than Rafale and SU-30 MKI when upgraded.

The PAF plans to operationally deploy the latest variant in 2020. It has initially ordered 50 Block IIIs to be delivered by 2024.[27] Older JF-17s may also be upgraded to the Block III variant later. The JF-17 Block II variant was known to cost around US$ 25 million. The new Block III will reportedly cost around $ 32 million each. The 26 two-seat JF-17B ordered will be used for training, and is a more effective electronic warfare platform with a second seat having a weapons systems officer (WSO) for operational role. The first eight JF-17B aircraft were produced in record time of just five months, and inducted by December 2019. The PAF expects to receive a further 14 JF-Bs in 2020 and four more in 2021.

Three JF-17's were sold to the Nigerian Air Force in 2018. At least six out of an order of 18 JF-17Ms have been delivered to Myanmar.[28] China and Pakistan are aggressively trying to find possible export customers. Targeted countries are Algeria, Argentina, Bangladesh, Egypt, Iran, Myanmar, Malaysia, Morocco, Nigeria, Sudan, Sri Lanka and Zimbabwe. The reasonable price makes it attractive.

JF 17B Two Seater

The JF-17B two-seater essentially incorporates a number of changes over the original JF-17 Block 2 fighter, including the addition of a second seat to include modified flight control systems, a three-axis fly-by-wire control system, an extended nose, a changed vertical stabiliser, altered material proportions and a greater wingspan.[29] The aircraft is also reported to house a more powerful radar than the original JF-17, though this remains to be confirmed, which would grant a greater degree of situational awareness. The fighter's twin seat configuration also allows it to serve as a trainer for future JF-17 pilots and will potentially allow it to serve as a more effective electronic warfare platform in which the second seat can accommodate a weapons systems officer (WSO) to carry out electronic attacks while the pilot flies the plane and deploys munitions. The PAF has inducted the JF-17B for both enhanced operational capability and training.

Tejas LCA vs JF-17

Comparisons are being drawn between the JF-17 and India's LCA. The Tejas uses many new technologies including large amounts of composite materials, advanced avionics and a unique aerodynamic configuration, and has a good potential to be expanded into variants. Some call the JF-17 the aircraft of today and the Tejas the aircraft of tomorrow. The JF-17 Block I costs close to $ 25 million,[30] vis-a-vis the LCA Mk 1 around US$ 23 million. The JF-17 Block II and III cost $ 28 million and $ 32 million, respectively, as per open sources. However, some publications say that the JF-17 Block II was sold to Myanmar for $ 16 million a piece.[31]

The LCA Tejas has been manufactured by a single country, and is claimed to be the world's lightest supersonic fighter. Currently only two squadron have been formed, with 22 aircraft. Aircraft production is still at below 8 a year, and is planned to be ramped up to 16. The more comparable LCA Mk 1A will have its first flight only in 2021 and inducted earliest by 2023. The Indian government has recently cleared 83 LCA MK1As to be purchased for Rs. 39,000 crore ($ 5.4 billion).[32] The deal for the upgraded single-engine fighters was expected to be signed by end 2020. The LCA Mk 1A will have its first flight only in 2021. Delivery of the Mark 1A will begin 36 months after the contract date. That would now mean 2023.

Debjit Sinha compares India's 'Tejas' LCA with the JF-17 in the context of selection by Malaysia for its Air Force.[33] Malaysia has short-listed India's indigenous LCA, JF-17 and F/A- 50 for procurement. For the first time ever, Tejas was showcased during the Langkawi International Maritime and Aerospace Exhibition (LIMA) in March 2019. Both Tejas and JF-17 were on display. Currently only two squadron with 22 aircraft have been formed. Aircraft production is still to be ramped up to 16 aircraft a year. The JF-17 is a joint project between China and Pakistan. Nearly 135 are already flying. Aircraft production at nearly 25 a year is high. The Block III will be inducted in 2020. There are already two foreign customers flying the JF-17. It has been in service for the last ten years and serves in six squadrons at full operational capability, whereas the Tejas has only two squadrons after four years. The JF-17's Russian engine has maintenance and serviceability issues well known to Malaysia from their MiG-29 experience. The LCA's General Electric F404 engine is much more reliable.

Another view is that cash-strapped Malaysia is now looking for lighter fighters to support Malaysia's F/A-18 and Su-30 fighters.[34] Pakistan's offer is the most affordable of the choices. Pakistan is offering the latest Block III variant. The 2021-2022 LCA program delivery timeframe means a Malaysian order could include some of the first fighters off the production line. Analyst and former Pakistan Air Force pilot Kaiser Tufail says that the JF-17 has been operational for the past 12 years and serves in six squadrons at full operational capability, whereas the Tejas was inducted just days ago "and has to go through the usual teething troubles." "On cost grounds, the JF-17 has a square chance," he said. Ben Ho, an air power analyst with the Military Studies Programme at Singapore's S. Rajaratnam School of International Studies, said the contenders have "fairly similar" performance, with individual advantages "marginal" or "negated in other areas." The JF-17 costs $25 million per unit, and the Tejas costs approximately $ 28 million. An order of 36 JF-17 fighters would mean that "a very substantial amount will be saved", Ho said. However, the JF-17's Russian engine may be problematic, as due to serviceability issues with the related engine of Malaysia's MiG-29s, potentially requiring "significant after-sales support and maintenance", Ho added. The Tejas is powered by the same General Electric F404 engine used in Malaysia's F/A-18s, and shares its weaponry with the Su-30s. However, the Tejas' Israeli avionics would likely need to be replaced, which "invariably means additional costs," he noted.

Sino-Pak Air Exercises

The PAF and PLAAF have participated in a series of exercises called Shaheen since 2011 to improve inter-operability to respond to 'mutual threats'. The missions have included simulated air combat, surface attack missions, air-refuelling and logistic support missions. Shaheen-I was held in Pakistan. Shaheen-II was held in September 2013 in Hotan in Western China. The PAF had then sent Mirage III EA and F-7G (MiG-21 class) aircraft. The PLAAF fielded J-10 multi-role fighters and J-7Cs. The more manoeuvrable J-10s acted as the aggressors. The three week-long Shaheen-III exercise was held in May 2014 at the PAF Rafiqi airbase near Shorkot in Western Punjab. The exercises gave both the Air Forces an opportunity to improve specific skills and to practice Dissimilar Air Combat Training (DACT). It also allowed them to train under different threat environment and training philosophies. The then PAF Deputy Chief of the Air Staff (Operations), Air Marshal Sohail Aman (later Air Chief), himself flew in an offensive air strike package. A six-member PLAAF delegation led by Major General Zhan Houshun, commander of Chengdu Regional Air Force Command, witnessed the exercise. He said "PLAAF was deeply impressed by PAF's aggressive combat style and brave fighting spirit and greatly moved by the streamlined efficient training approach, and hoped the exercise would deepen and broaden friendship and cooperation between the two Air Forces."[35] The exercise was also designed to learn modern air warfare and share operational thoughts among pilots, radar controllers. These exercises were of special importance to the PAF as it gave them exposure to fly against Chinese Sukhoi Su-27/Su-30MKK aircraft which are similar to Indian Air Force frontline SU-30 MKI aircraft and to help them validate their tactics.

The most recent Shaheen VIII[36] (Eagle VIII) was also held at Hotan in south-western Xinjiang, north east of the Indian town of Leh in Ladakh in August 2019. The air exercises are said to be an exercise to develop a mechanism for the interoperability of the air forces of both countries. Pakistani JF-17 aircraft transited through the Skardu airbase in Gilgit Baltistan to the exercise location. Gilgit Baltistan, a semi-provincial area occupied by Pakistan, which India claims is a part of the former princely state of Jammu and Kashmir and is an integral part of the country. The Pakistanis used the base after a long time to join the Chinese Air Force which has built several bases in the Tibetan Autonomous Region (TAR) not far from the border with India. This was the first war-game after the Abrogation of Article 370. Islamabad is participating with its JF-17 fighter aircraft while China is taking part with its J-10 and J-11 fighters. The PAF has over a 120 JF-17 'Thunder'. On the other hand, the Chinese J-10 and J-11 form the modern backbone the PLAAF fighter fleet. The latest variant J-10C has an active electronically scanned array (AESA) radar and the ability to carry PL-10 and PL-15 air-to-air missiles. Further details of the equipment used and whether battle training was also part of the exercise are not yet clear. The joint training between the air forces of the all-weather friends included contingents comprising combat pilots, air defence controllers, and technical ground crew along with fighter jets, bombers and early warning AWACS planes, according to a report in Pakistan's Geo TV.[37] The PAF so far has held operational exercises with air forces of the USA, Turkey, China, Saudi Arabia, UAE, Jordan, Egypt and a few others.

Russian Interest in Pakistan

With India becoming closer towards the USA, Russia and Pakistan signed their very first military cooperation agreement in 2014 and laid out future avenues of cooperation, ending years of the situation wherein Islamabad remained closer to the USA and Moscow to India.[38] Joint naval exercises, military officers exchange, arms sales, counter-narcotics and counter-terrorism cooperation are envisaged between Russia and Pakistan. The sale of 20 Mi-35 transport helicopters to Pakistan, and Pakistani interest in the Su-35 are being talked about. Russia wants to be active in the region when the USA withdraws troops from Afghanistan. Yet Russia's interest in India dominates. Russia accounted for 58 per cent of total Indian arms imports in 2014-2018, down from 76 per cent between 2009 and 2013. But this figure will once again go up after some big ticket sales such as the S-400, among others. Pakistan is at the 11th position in defence imports accounting for 2.7 per cent of all global arms imports. It imports 70 per cent from China, 8.9 per cent from the USA and interestingly 6 per cent from Russia.

IMPLICATIONS AND OPTIONS INDIA

China's Desire to Dominate

China's desire to dominate Asia is the first step to finally dominate the world. They would prefer a unipolar Asia and bi-polar world. Over the last few decades, China has very systematically been working to encircle and contain India, the only possible challenger. It is also to increase its sphere of influence in Indian Ocean Region, and in turn prevent the USA (along with friends) to encircle China. Pakistan is a critical link in this plan. It has been overtly/covertly helping it build Pakistan's aircraft industry, missile technology and nuclear power among others. The China-built-and-operated strategic Gwadar port near the Iranian border in otherwise turbulent Baluchistan will become significant when the USA pulls out of Afghanistan. The aggressive expansion of PLAN is part of this grand strategy. China and Pakistan are deepening and broadening military cooperation in such fields as joint exercise and training, so as to enrich the strategic cooperation between the two countries, said Chinese President Xi Jinping during a visit to Pakistan in April 2015. Hailing the important role the Pakistani military has played in the development of China-Pakistan relations, Xi said that the understanding and support of the Pakistani military is necessary for the two countries to forge a community of shared destiny.[39] Pakistani military leaders affirmed their support for the consolidation of the friendly cooperation between China and Pakistan, and staunchly back China's efforts to counter terrorism. Earlier, Pakistan escorted the Chinese President Xi's plane with eight fighter escorts as a mark of the military's respect and gratitude.[40] Pakistan believes that America's influence and support in the region should be counterbalanced by the Chinese.

During Pakistan PM Imran Khan's third trip to Beijing in less than a year, in September 2019, Chinese President Xi Jinping offered his support to Pakistan on the issue of Kashmir and boosted military ties between the two just days before a meeting with Indian Prime Minister Narendra Modi.[41] Khan "underscored that China was Pakistan's steadfast ally, staunch partner and iron brother" during the meeting with Xi. Khan and "maintained that China had firmly stood by Pakistan in supporting Pakistan's core national interests and played a major role in advancing Pakistan's economic as well as national development goals."

At $ 177.6 billion, China's 2019 defence budget is nearly four times that of India's US$ 44 billion. With the U.S. focus, China is clearly pulling ahead of India in defence technology and capability. India-China bilateral trade is nearly US$ 100 billion. China-Pakistan trade on the other hand is around US$ 10 billion. With the USA and the Europeans tightening imports from China, it needs Indian markets. Also, the Wuhan informal summit had built mutual confidence measures. Pakistan is heavily dependent on Chinese debt in CPEC. Pakistan's Gwadar Port, built and managed by China, has run into rough weather due to inadequate cargo handled at the port and lack of use of its use for transit to Afghanistan casting a shadow over Beijing's BRI in the region. Pakistan's foreign exchange reserves are at an all-time low. There is a risk of it defaulting and is currently looking for financial aid from friends. Pakistan is also facing the threat of being put on the Financial Action Task Force's (FATF) blacklist. Despite strong political rhetoric at the UNGA, under these circumstances, Pakistan is in no position to threaten India.

China's first focus is to emerge as the leading global power. Its stated first aim is Taiwan and dominancy on its eastern seaboard. It wants stability on other borders. Despite hard stand in Galwan, Ladakh, China is unlikely to start a full war in at least the next decade. In the past Indo-Pak conflicts, China has not intervened militarily. At best, it will provide military logistics support. In case of an unlikely event of even a short Sino-Indian skirmish, Pakistan is likely to take advantage and open a front.

Options and Approach India

The Indian armed forces have been factoring a two-front war in their plans and have been war gaming and testing the same in major exercises. It is an important part of capability building. The IAF carried out a major exercise, 'Gagan Shakti', in 2018 to simulate a two-front war with existing combat assets and fighter squadrons which are at an all-time low of 29. In case of such a conflict, the Indian approach would have to be to gain ground against Pakistan and dissuade and deter China by holding ground firmly. India is building military assets in the north east. All Advanced Landing Grounds have been modernized. Airfield infrastructure including hardened shelters are being built. SU-30 MKIs are already positioned. Newer air assets such as the Rafale, C 130 J, Chinook and Apache helicopters are being positioned at eastern airbases.

Aerospace Options for India

China has a sizable air power when the air assets of the PLAAF and PLAN are combined. The PAF has plans to increase its current 20 squadrons to its target of 26. The Current IAF : PAF ratio of 1.4:1 is a far cry from the once 3:1 dominance. The IAF is currently at an all-time low of 29 fighter squadrons. India thus needs to re-look at the force structure. Most analysts believe that for a two-front war, 45 to 50 fighter squadrons are required. India currently has only three IL-76 'Phalcon' platform AWACS. Two more are in the pipeline. At least two more are required. The IAF also has two ERJ 145 DRDO 'Netra' Airborne Early Warning and Control (AEW&C) systems. These are inadequate; at least ten of them are required in the long term. The current six FRA must go up to 12. Acquisition process for the same has been mired by delays.

To achieve this, the defence budget must be increased to at least 2.5 per cent of the GDP from the current 1.44 per cent. Based on the Rafale contract, it is clear that typically a squadron of modern fighters costs US$ 4.3 billion. The IAF has to make good the 12 squadrons short. Besides, another nine are due to be phased out in the next decade. Thus, 21 new squadrons will require US$ 90 billion at current prices. These are large sums. The IAF's budget share is Rs. 68,949 crore, of which capital outlay is Rs. 39,347 crore (US$ 5.5 billion). Large committed liabilities for past purchases are currently taking away all the capital budget. The IAF will thus require out-of-budget funds for purchases. Also, India needs to hasten procurement processes. The Defence R&D and Indian aircraft industry too would have to get their act right if Make-in-India has to succeed. There is a need for the IAF to build up force levels quickly lest it gets left too far behind.

The IAF immediately requires advanced fighters, sophisticated support platforms and smart long-range weapons. The Defence R&D and Indian aircraft industry too would have to get their act right. The IAF's long list of acquisitions of 114 medium multi-role fighters, the hastening of development and production of the LCA and AMCA, early finalization of replacement of AVRO aircraft and the various other force multipliers need to unfold quickly. The Defence Procurement Policy (DPP) 2016 needs further review to support the Make-in-India drive and encourage India's private industry in a bigger way. FDI needs to increase. India needs to have at least two more major operational airfields in the Assam valley and deploy more Surface-to-Air Missiles. There is a need for the IAF to build up force levels quickly lest it gets left too far behind the PLAAF and PAF combine. National security cannot be left open-ended.

REFERENCES

1. Brainy Quote attributed to Pak Politician Shehbaz Sharif, https://www.brainyquote.com/quotes/shehbaz_sharif_830230
2. China says Pakistan 'most reliable friend', *Dawn Newspaper*, January 26, 2015, https://www.dawn.com/news/1159585
3. Sushant Sareen, ORF Raisina Debates, February 20, 2019, https://www.orfonline.org/expert-speak/pakistan-china-new-america-48305/
4. CIA 'Operation Cyclone', OER Services, https://courses.lumenlearning.com/suny-hccc-worldhistory2/chapter/the-united-states-and-the-mujahideen/
5. Rabia Akhtar, Pressler Amendment, The correct narrative on Pressler, dawn.com, May 29, 2017, https://www.dawn.com/news/1335979
6. Bilateral Relations: India-Russia Relations, Embassy of India, Moscow, https://indianembassy-moscow.gov.in/bilateral-relations-india-russia.php
7. Dinakar Peri, What is LEMOA?, *The Hindu*, October 18, 2016, https://www.thehindu.com/news/national/What-is-LEMOA/article15604647.ece
8. Seven reasons why COMCASA is so important for India, *Economic Times*, September 07, 2018, https://economictimes.indiatimes.com/news/defence/seven-reasons-why-comcasa-is-so-important-for-india/articleshow/65707682.cms?from=mdr
9. Wenwen Shen, China and its Neighbours: troubled relations. EU-Asia Centre, March 01, 2012, http://www.eu-asiacentre.eu/pub_details.php?pub_id=46
10. Maninder Dabas, Here Is All You Should Know About 'String of Pearls', *India Times*, June 23, 2017, https://www.indiatimes.com/news/india/here-is-all-you-should-know-about-string-of-pearls-china-s-policy-to-encircle-india-324315.html
11. Catherine Wong, China-Pakistan military ties, *South China Morning Post*, January 07, 2018, https://www.scmp.com/news/china/diplomacy-defence/article/2127106/china-pakistan-military-ties-set-get-even-closer-iron
12. Ambuj Sahu, China-Pakistan arms transfer, ORF, June 14, 2019, https://www.orfonline.org/expert-speak/analysing-trends-arms-transfer-china-pakistan/
13. Sutirtho Patranobis, China-Pak relations 'higher than mountain, deeper than ocean', *Hindustan Times*, May 23, 2013, https://www.hindustantimes.com/world/china-pak-relations-higher-than-mountain-deeper-than-ocean/story-22JlWjJtaqGWZJ6okMwzXK.html
14. Pakistan-China Relation Major developments, https://shodhganga.inflibnet.ac.in/bitstream/10603/208927/11/14-appendices.pdf
15. J-10 (Jian 10) Vigorous Dragon Multirole Tactical Fighter, Thai Military and Asian Region, https://thaimilitaryandasianregion.wordpress.com/2015/10/25/chengdu-j-10/
16. Sebastein Roblin, Meet the JF-17 Block III Fighter: The Jet China is Helping Pakistan Build to Fight India, *The National Interest*, August 10, 2019, https://nationalinterest.org/blog/buzz/meet-jf-17-block-iii-fighter-jet-china-helping-pakistan-build-fight-india-72751
17. PTI, China to sell 48 high-end military drones to Pakistan, October 10, 2018, https://economictimes.indiatimes.com/news/defence/china-to-sell-48-high-end-military-drones-to-pakistan/articleshow/66129500.cms?from=mdr

18. Shaanxi Y-8 Military Transport Aircraft, Air Force Technology, https://www.airforce-technology.com/projects/shaanxiy8militarytra/
19. K-8 Karakorum Light Attack and Jet Trainer Aircraft, Air Force Technology, https://www.airforce-technology.com/projects/hongdu-k8/
20. Frenz-Stefan Gady, Pakistan to Stick With Russian Engine for JF-17 Fighter Jet, *The Diplomat*, November 25, 2015, https://thediplomat.com/2015/11/pakistan-to-stick-with-russian-engine-for-jf-17-fighter-jet/
21. Putin Clears sale of RD-93 fighter jet engines by China to Pakistan. domain-b.com. April 27, 2007, https://www.domain-b.com/aero/20070427_putin.htm
22. Jon Lake, More 'Thunder' in China's Air, AINonline, January 02. 2020, https://www.ainonline.com/aviation-news/defense/2020-01-02/more-thunder-chinas-air
23. Flying and Fighting in the JF-17 Thunder: Interview with a PAF Fighter Pilot, Hush Kit, July 19, 2019, https://hushkit.net/2019/07/19/flying-fighting-in-the-jf-17-thunder-interview-with-pakistan-air-force-fighter-pilot/
24. Usman Ansari, What does Pakistan need to close its air defence gaps? *Defence News*, April 10, 2019, https://www.defensenews.com/global/asia-pacific/2019/04/10/what-does-pakistan-need-to-close-its-air-defense-gaps/
25. Jeffrey Lin and P.W. Singer, Chinese Air-To-Air Missile Hits Targets, Spooks USAF General, *Popular Science*, September 22, 2015, https://www.popsci.com/chinese-air-to-air-missile-hits-targets-spooks-usaf-general/
26. Frenz-Stefan Gady, Pakistan's JF-17 Block III Fighter Jet Makes Maiden Flight, *The Diplomat*, January 06, 2020, https://thediplomat.com/2020/01/pakistans-jf-17-block-iii-fighter-jet-makes-maiden-flight/
27. Frenz-Stefan Gady, Pakistan's JF-17 Block III Fighter Jet to Make Maiden Flight by the End of 2019, *The Diplomat*, July 11, 2019, https://thediplomat.com/2019/07/pakistans-jf-17-block-iii-fighter-jet-to-make-maiden-flight-by-the-end-of-2019/
28. Sebastein Roblin, No Stealth Jets For You: How Pakistan Is Making Due With The JF-17 Thunder, *The National Interest*, December 14, 2019, https://nationalinterest.org/blog/buzz/no-stealth-jets-you-how-pakistan-making-due-jf-17-thunder-104802
29. Military Watch, Myanmar Air Force Receives Twin Seat JF-17B Fighter Amid Air Force Modernisation Efforts, March 16, 2019, https://militarywatchmagazine.com/article/myanmar-air-force-receives-twin-seat-jf-17b-fighters-amid-air-force-modernisation-efforts
30. Tom Demerly, Shopping for Fighters: Is the Chinese/Pakistani JF-17 Thunder the Real "Joint Strike Fighter"? *The Aviationist*, November 09, 2017, https://theaviationist.com/2017/11/09/shopping-for-fighters-is-the-chinesepakistani-jf-17-thunder-the-real-joint-strike-fighter/
31. Charlie Gao, China's JF-17 Multirole Jet Could Be A Big Success, *The National Interest*, January 01, 2020, https://nationalinterest.org/blog/buzz/chinas-jf-17-multirole-jet-could-be-big-success-thanks-f-16-dna-110111
32. Rahul Bedi, Indian MoD concludes price negotiations with HAL for 83 LCA Mk 1A combat aircraft, *Jane's Defence Weekly*, February 18, 2020, https://www.janes.com/article/94369/indian-mod-concludes-price-negotiations-with-hal-for-83-lca-mk-1a-combat-aircraft
33. Debjit Sinha, India's Tejas vs China-Pakistan's JF-17! Malaysia to pick one for its Air Force, *Financial Express*, March 25, 2019, https://www.financialexpress.com/defence/tejas-vs-jf-17-2019-indian-air-force-pakistan-lima-malaysia/1526975/
34. Usman Ansari, How does Pakistan's Thunder fare against contenders in Malaysia's

aircraft competition? *Defence News*, February 21, 2019, https://www.defensenews.com/air/2019/02/21/how-does-pakistans-thunder-fare-against-contenders-in-malaysias-aircraft-competition/

35. Staff Reporter, Pak-China joint air drill concludes, *The Nation*, May 28, 2014, https://nation.com.pk/28-May-2014/pak-china-joint-air-drill-concludes
36. All you need to know about Shaheen VIII military exercise, *Firstpost*, August 27, 2019, https://www.firstpost.com/india/pakistan-china-engage-in-joint-air-force-exercise-300-km-north-of-leh-all-you-need-to-know-about-shaheen-viii-military-exercise-7232121.html
37. Geo News, Pakistan, China air forces start joint exercise Shaheen VIII, Aug 24 2019, https://www.geo.tv/latest/245710-pakistan-china
38. Matthew Bodner, Russia Signs Landmark Military Cooperation Agreement With Pakistan, *The Moscow Times*, November 20, 2014, https://www.themoscowtimes.com/2014/11/20/russia-signs-landmark-military-cooperation-agreement-with-pakistan-a41589
39. Xinhua, China, Pakistan to deepen military cooperation, Chinese MoD, April 21, 2015, http://eng.mod.gov.cn/DefenseNews/2015-04/21/content_4581065.htm
40. Zhao Lei, Chinese-made fighter jets escort Xi, *China Daily*, April 22, 2015, https://www.chinadaily.com.cn/world/2015xivisitpse/2015-04/22/content_20503724.htm
41. Tom C'Connor, China's Xi supports Pakistan on Kashmir, boosts military ties ahead of meeting with India, *Newsweek*, September 10, 2019, https://www.newsweek.com/china-support-pakistan-military-india-1464269

9

Strategic Balance in the Indo-Pacific

"Only air power can defeat air power. The actual elimination or even stalemating of an attacking air force can be achieved only by a superior air force."

—Alexander P. de Seversky

"Indo-Pacific is a region with 50 per cent of the world's population, and enormous diversity of religions, culture, languages, history and political and economic systems."[1]

—Prime Minister Narendra Modi

Asia-Pacific typically includes much of East Asia, South Asia, Southeast Asia, and Oceania. It may also include parts of Russia (on the North Pacific) and countries in the Americas which are on the coast of the Eastern Pacific Ocean. Overall, there is no clear-cut definition of "Asia Pacific", and the regions included change depending on the context. The Indo-Pacific, on the other hand is a bio-geographic region of the Earth's seas, comprising the tropical waters of the Indian Ocean, and the western and central Pacific Ocean, and the seas connecting the two in the general area of Indonesia. It does not include the temperate and polar regions of the Indian and Pacific oceans, nor the Tropical Eastern Pacific. Initially used by Indian and Australian scholars, the term was made prominent by Japan's Prime Minister Shinzo Abe, as reflected in his speech to the Indian Parliament in August 2007[2] that talked of the "Confluence of the Indian and Pacific Oceans" as "the dynamic coupling as seas of freedom and of prosperity" in the "broader Asia". On the eve of the Shangri La Dialogue at Singapore in June 2018, the U.S. Defence Secretary Jim Mattis declared that the U.S. Pacific Command would henceforth be called the U.S. Indo-Pacific Command. This was not an off-the-cuff remark but the culmination of a series of events that preceded the eventual declaration.[3]

Indo-Pacific: The Term Evolves

Since 2011, the term 'Indo-Pacific' is being used increasingly in the global strategic/geopolitical discourse. The German geo-politician Karl Haushofer first used the term in the 1920s. Since then, many analysts sought to describe the 'geo-economic' connect between the Indian and Pacific oceans. However, in the contemporary context, beginning in the 2000s, analysts began to observe the 'security' linkage between the two oceans. The term was first used in an article authored by Gurpreet Khurana, which was carried in the January 2007 issue of the *Strategic Analysis Journal* (Routledge/IDSA) titled "Security of Sea Lines: Prospects for India-Japan Cooperation."[4] In the article, the term 'Indo-Pacific' refers to the maritime space stretching from the littorals of East Africa and West Asia, across the Indian Ocean and western Pacific Ocean, to the littorals of East Asia. From 2010 onwards, the term 'Indo-Pacific' acquired salience within the Indian government and has since been used often by India's apex political leadership. From about 2011 onwards, the term has been used frequently by strategic analysts and high-level government/military leadership in Australia, Japan and the USA to denote the said region. However, a formal/official documented articulation of the term first appeared in Australia's Defence White Paper, 2013.[5] It has been argued that the concept of the Indo-Pacific may lead to a change in popular "mental maps" of how the world is understood in strategic terms. Since 2013, US officials have begun using the term "Indo-Asia Pacific". This enabled America to maintain its geographic inclusiveness in the new coinage of 'Indo-Pacific'. With the rising involvement of the USA in the new growth areas of Asia, the idea of the Indo-Pacific Economic Corridor was conceptualized during the US-India Strategic Dialogue of 2013, where Secretary of State John Kerry referred to the potential of the Indo-Pacific Economic Corridor in transforming the prospects for development and investments as well as for trade and transit between the economies of South and Southeast Asia Indo-Pacific economic corridor.

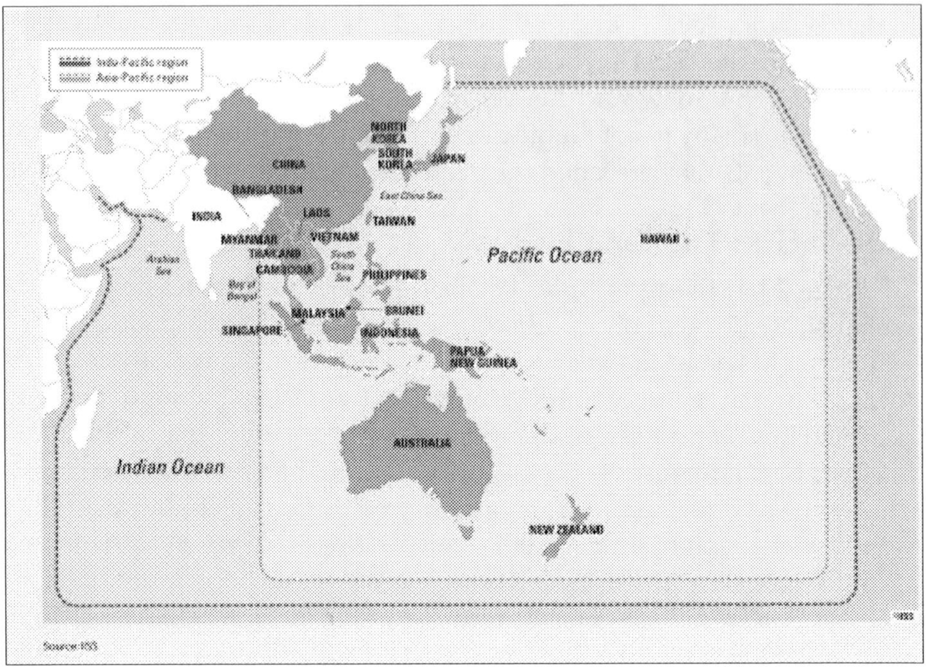

Asia-Pacific and Indo-Pacific.
Image Source: IISS

The term's profile was raised when it found mention in the joint statement issued by the Indian Prime Minister Narendra Modi and the United States President Donald Trump after the former's state visit to the White House on 26 June 2017.[6] "As responsible stewards in the Indo-Pacific region, President Trump and Prime Minister Modi agreed that a close partnership between the USA and India was central to peace and stability in the region. In marking 70 years of diplomatic relations between India and the USA, the leaders resolved to expand and deepen the strategic partnership between the countries and advance common objectives. Above all, these objectives include combating terrorist threats, promoting stability across the Indo-Pacific region, increasing free and fair trade, and strengthening energy linkages".

Power shifts in the Indo-Pacific

The Australian Government's Foreign Policy White Paper of 2017 indicates that the economic growth in Asia continues to re-shape the strategic landscape.[7] The compounding effect of China's growth is accelerating shifts in relative economic and strategic weight. In parts of the Indo–Pacific, including in Southeast Asia, China's power and influence are growing to match, and in some cases exceed, that of the USA. The future balance of power in the Indo-Pacific will largely depend on the actions of the USA, China and major powers such as Japan and India. The responses of major Southeast Asian states, such as Indonesia and Vietnam, will also be important.

China, which became the world's largest economy in purchasing-power-parity terms in 2014, is a global power in scale but not always in global integration. It became the world's largest trading nation of goods in 2013.[8] However, although China has 110 Global Fortune 500 companies, more than 80 per cent of its revenue is still earned at home. China is banking, securities, and bond markets rank in the global top three in size, but international players have a limited presence. The relationship between China and the world is changing. China is the most important trading partner for most of the region's economies and a major investor, including in infrastructure. China's military modernization is rapidly improving the capability of its armed forces. It has the largest navy and air force in Asia and the largest coast guard in the world. It is a large aid donor and lender to the region.

Like all great powers, China will seek to influence the region to suit its own interests. As it does, a number of factors suggest we will face an increasingly complex and contested Indo–Pacific.[9] Even as China's power grows and it competes more directly with the USA regionally and globally, the USA will, for the near future, retain its significant global lead in military and soft power. It will continue to be the wealthiest country in the world, measured in net asset terms, the world's leader in technology and innovation, and home to the world's deepest financial markets.

The United States' long-term interests will anchor its economic and security engagement in the Indo-Pacific.[10] Its major Pacific alliances with Japan, the Republic of Korea and Australia will remain strong. Japan and India, major economies and military powers in their own right, are also playing stronger roles in Indo-Pacific security and political affairs and are seeking to influence the balance of the regional order. In this dynamic environment, competition is intensifying. Maritime and land border disputes will continue to create friction. The region's seas and airspace are becoming more contested. Freedom of navigation is under challenge in parts of the region. Economic power is also being used for strategic ends, including the financing of infrastructure projects.

India and the Indo-Pacific Balance of Power

Nitin Pai, Director, The Takshashila Institution, wrote that ultimately, India's most important role in the Asia-Pacific is to be a more successful form of itself: demonstrating that strong economic growth can be achieved within a diverse, plural, liberal democracy.[11] Over the next few decades, as China closes in on the USA as a world power, India finds itself emerging as a swing power. Its economic, military and diplomatic power affords it the ability to influence, albeit not yet decisively, the contest between the world's two biggest powers. Even as they improved relations with China and the USA, Indian governments since the end of the Cold War have scrupulously resisted allowing their engagement with one be seen as being directed against the other, he writes. As Kissinger observed, being a effective swing power requires India to enjoy better relations with the USA and China; to have the capacity to both benefit and impose costs on the bigger powers; and finally, political and diplomatic dexterity to take positions issue-by-issue. While the Indian government has not officially enunciated such a doctrine, its actions have generally been in this direction. New Delhi has pursued a strategic partnership with the USA since the early 2000s, and despite a longstanding border dispute, participated in Chinese-led initiatives like the SCO and the Asia Infrastructure Investment Bank (AIIB). Washington has systematically courted New Delhi, particularly in matters concerning the Indo-Pacific region, even while attempting to manage differences over the Middle East, Iran, Afghanistan and Pakistan. The signing of the India-US nuclear agreement in 2005 and the subsequent mainstreaming of India in the Nuclear Suppliers Group was a landmark in this regard. Since then, there is an ever-closer mutually bipartisan consensus on widening and deepening the India-US relationship. China's backing of Pakistan and shielding of Pakistan-based international terrorists from UN sanctions, is actually pushing India into a deeper security relationship with the USA.

India's Look East Policy

India's Look East Policy is an effort being made by the Indian government to cultivate and strengthen economic and strategic relations with the nations of Southeast Asia in order to solidify its standing as a regional power.[12] It also serves to position India as a counterweight to the strategic influence of China in the region. Initiated in 1991, it marked a strategic shift in India's perspective of the world. It was developed and enacted during the government of Prime Minister P.V. Narasimha Rao and has continued to enjoy energetic support from the successive administrations of different political parties in India. Only after China started fostering trade and economic relations with other Asian nations did India realise the need to secure its eastern flank by making inroads and to act as a counter balancing force. According to former Indian Ambassador Rajiv Sikri, India missed a crucial opportunity during this period to leverage India's shared colonial experience, cultural affinities and lack of historical baggage to build strong economic and strategic relations with Southeast Asia.[13]

Much of India's Look East Policy involves Myanmar, which is the only Southeast Asian country that shares a border with India and is seen as India's gateway to Southeast Asia. The South Asia Free Trade Agreement created a free trade area of 1.6 billion people including Bangladesh, Bhutan, Maldives, Pakistan and Sri Lanka in 2006. The ASEAN–India Free Trade Area (AIFTA), with ten members came into effect in 2010. India also has separate free trade agreements with Sri Lanka, Japan, South Korea, Singapore, Thailand and Malaysia, among others.

India has also boosted its cooperation with Asian regional groupings such as ASEAN, the Bay of Bengal Initiative for Multi-Sectoral Technical and Economic Cooperation (BIMSTEC) and the South Asian Association for Regional Cooperation (SAARC). Indian companies have made significant economic and trade agreements in infrastructure and other areas. Some of the major projects taken up by India include the resurfacing and upgrading of the 160-kilometre Tamu-Kalewa-Kalemyo road and the Kaladan project that will connect Kolkata Port with Sittwe Port in Myanmar (still in progress). Once these infrastructure projects are completed, the next step will be connecting the India-Myanmar highway network to the existing portions of the Asian Highway Network, which will connect India to Thailand and the rest of Southeast Asia. Indian policy attempts to connect the Indian economy to its traditional maritime neighbourhood and trading partners, broadly across both sides of the Straits of Malacca. While the shifting of the global balance towards the east, resulting in new multilateral international financial institutions, can be in India's interests, New Delhi remains concerned that China's regional dominance should not unfavourably change the norms.

Belt and Road Initiative: Indian Concern

India rightly fears that China's BRI is essentially a Chinese Belt and Chinese Road. Relatively poor India's Asian neighbours are looking at low-cost Chinese financing and technical assistance for better connectivity, but in effect getting into a debt-trap that could one day force them to give much land access to China as has happened in Pakistan and Sri Lanka, among others. This presence of China is to the detriment of India's geo-strategic interests in the region. Senior Indian officials have expressed concern over the purpose of Chinese investment in infrastructure in regions where there is no discernible commercial purpose.[14] Having not received satisfactory answers from their Chinese interlocutors, New Delhi is apprehensive of the risk of military facilities being surreptitiously built in its proximate and broader maritime neighbourhood. When Indian commentators say China is constructing a "string of pearls" in the Indian Ocean, the unexpressed worry is that of being contained and losing preponderance in its immediate maritime neighbourhood.

India's Balancing Act

India's primary objective in the Indo-Pacific is to prevent China from dominating the region. In a major foreign policy speech at the Shangri-La Dialogue in June 2018, Prime Minister Narendra Modi repeatedly invoked the principle of the equality of all nations, 'large and small', a not-too-subtle reference to China's relations with its smaller neighbours. 'We see the assertion of power over recourse to international norms,' he said, in another reference to China's assertive behaviour in its neighbourhood, and called for a 'common rules-based order for the region ... that must apply to all ... based on sovereignty and territorial integrity'.[15] While these are not unusual principles to invoke, doing so in the context of the Indo-Pacific cannot but be read as an implicit statement that India does not accept that any special privileges should accrue to China, despite the gross disparity between Chinese power and that of other states in the region.

It is not just India that is attempting to fashion a complicated mix of policies to deal with China's rise and its consequences for the region, and finding the process fraught with contradiction. Most of China's other neighbours are also engaged in similar efforts. Paradoxically, although Prime Minister Modi declared that 'India does not see the Indo-Pacific as a strategy', the strategic elements of the country's Indo-Pacific policy make it less likely that Beijing will believe him, and thus even less likely that this reassurance strategy will work.[16] Thus, while 'Look East' and 'Act East' represented, at least formally, an effort to foster closer cooperation with Asian economies to its east, India's approach to the Indo-Pacific is much more directly designed to counter China's growing power and expansion into South Asia and the Indian Ocean.

India's Emerging Role

India has begun as a contributor to the Indo-Pacific balance, through joint military exercises, patrolling, port calls, anti-piracy missions and humanitarian missions.[17] The Indian Navy has increased the frequency of its engagements east of the Straits of Malacca. New Delhi's longstanding reluctance to participate in multilateral military exercises might not endure if Beijing continues on its current, antagonistic trajectory. India and Japan are moving closer with economic partnership, and Japan has offered the US-2 amphibious aircraft to India. Most ASEAN nations have their own axe to grind with China and some may be sceptical to antagonise China and may be doing fine balancing. Ultimately, what will count for India is its own economic and military strength for prestige and influence in the region.

2017 Doklam Standoff

The 2017 China-India Doklam border standoff occurred over Chinese construction of a road in Doklam near a tri-junction border area. On June 16, 2017, Chinese troops with construction vehicles and road-building equipment began extending an existing road southward in Doklam, a territory which is claimed by both China, and India's ally, Bhutan. On June 18, 2017, as part of Operation Juniper, about 270 Indian troops armed with weapons and two bulldozers crossed the Sikkim border into Doklam to stop the Chinese troops from constructing the road. On August 28, after nearly 75 days, both India and China announced that they had withdrawn all their troops from the face-off site in Doklam. The Ministry of External Affairs (MEA) of India released a press statement stating that India and China had mutually agreed to disengage. In Beijing, the foreign ministry spokesperson said that the Chinese forces on site have verified that the Indian troops pulled out, and implied that Chinese troop numbers would be reduced. She said that the Chinese troops would continue to patrol the area, to garrison it and to exercise "sovereign rights". However, she made no mention of road-building activities. The statement offered Beijing a face-saving way out of the impasse. This stand-off, closely watched by Asian countries, especially those who have territorial and maritime disputes with China, has shown that China's expansionist ambition is not unstoppable.[18] While Japan openly supported India's position in Doklam and in the cases of Vietnam and other SE Asian nations, which have been victims of Beijing's ambitions in the South China Sea region, it monitored the situation from close quarters. India's strategic restraint not only restored the status quo but also enabled it to enhance Delhi's profile in the comity of nations as an emerging power. During Chinese Foreign Minister Wang Yi's visit to India in December 2017, he said that despite the Doklam dispute which had complicated relations, China and India have far greater shared strategic interests than differences.[19] While gently worded, Wang's remarks do not eradicate from the collective Indian memory the vitriol about teaching India "a bitter lesson" that stemmed from Chinese state media during the Doklam standoff. India's resistance to China's Belt and Road Initiative, a project that is clearly viewed as an influence-building initiative, also brought a level of chill in the relationship. The BRI is viewed by India as exclusionary to Indian economic and strategic interests in the region and beyond.

Trans-Pacific Partnership (TPP)[20]

The Trans-Pacific Partnership was a proposed trade agreement between Australia, Brunei, Canada, Chile, Japan, Malaysia, Mexico, New Zealand, Peru, Singapore, Vietnam, and the USA signed on 4 February 2016, which was not ratified as required and did not come into effect. After the newly-elected US President Donald Trump withdrew the US signature from TPP in January 2017, the agreement could not enter into force. The remaining countries negotiated a new trade agreement called Comprehensive and Progressive Agreement for Trans-Pacific Partnership, which incorporates most of the provisions of the TPP and which came into force on 30 December 2018. Many observers have argued that the trade deal would have served a geopolitical purpose, namely, to reduce the signatories' dependence on Chinese trade and bring the signatories closer to the USA. The TPP agreement was part of President Obama's pivot. Meanwhile China continues to make its own trade deals with key nations in the region, demonstrating a serious decline in American hard and soft regional power.

Quadrilateral Security Dialogue (Quad)

US Indo-Pacific strategy reflects three processes for the USA. The first is that China's rise presents a "power transition" challenge for the USA. The second is that the USA is responding to that challenge by pursuing "balancing", both in terms of building up its own strength ("internal balancing") and in strengthening its alliances and strategic partnerships ("external balancing"). The third is that in balancing terms, "balance of threat" considerations are in operation not only for the USA but also for Japan, Australia and India vis-à-vis China.[21] The USA welcomes India's rise because it provides an increasingly important balancing counterweight to China's rise. Japan and India together could be able to balance with China. Perceived offensive intentions and geographic proximity make India and Japan, among a few others, more concerned about Chinese power than about US power.

The Quad is an informal strategic dialogue between the USA, Japan, Australia and India that is maintained by talks between member countries. The dialogue was initiated in 2007 by Prime Minister Shinzo Abe of Japan. The dialogue was followed by a joint military exercise called 'Exercise Malabar'. The diplomatic and military arrangement was widely viewed as a response to increased Chinese economic and military power. The Chinese government responded to the Quadrilateral dialogue by issuing formal diplomatic protests to its members. The significance of Quad increased after the tensions caused by Chinese territorial ambitions in the SCS. As Indian Foreign Minister S. Jai Shankar engaged with his counterparts from BRICS as well as USA-Japan-Australia in Quad format in September 2019, in what was almost back-to-back meets, the U.S. Acting Assistant Secretary for the Bureau of South and Central Asian Affairs, Alice Wells, told media in New York[22] "a wide-ranging discussion of our collective efforts to advance a free and open Indo-Pacific, but also touching on counterterrorism, humanitarian assistance and disaster relief, maritime security cooperation, development finance, and cyber-security. I would say that the U.S. and Indian joint participation in the Quad also demonstrates the strength of the U.S.-India relationship and our shared commitment to, again, advancing a values-based policy towards the region."

US-Indian Military Cooperation

Active US-Indian military cooperation expanded in 1991 following

India's the economic liberalisation when the USA proposed army-to-army cooperation. A "New Framework for India-US Defence" was signed in 2005, increasing cooperation regarding military relations, the defence industry and technology sharing, and the establishment of a "Framework on maritime security cooperation". India and the USA have conducted dozens of joint military exercises since then before the development of Quad dialogue. including Cope-India (Air Force), Yudh Abhyas (Army) and Vajra Prahar (Special Forces). The two sides are also increasingly engaged in multi-lateral exercises such as the Malabar, Red Flag, and RIMPAC, covering the broad expanse of the Indo-Pacific.[23]

The USA has four "foundational" agreements that it signs with its defence partners. With India the General Security of Military Information Agreement (GSOMIA), was signed in 2002. The agreement enables the sharing of military intelligence between the two countries. The Logistics Exchange Memorandum of Agreement (LEMOA), was signed on 29 August 2016. It permits the military of either country to use the other's bases for re-supplying or carrying out repairs. The Communications Compatibility and Security Agreement (COMCASA) was signed during the inaugural 2+2 dialogue in September 2018. It is an India-specific variant of Communications and Information Security Memorandum of Agreement (CISMOA) that enables the two countries to share secure communication and exchange information on approved equipment during bilateral and multinational training exercises and operations. The Basic Exchange and Cooperation Agreement (BECA) has not yet been signed. It permits the exchange of unclassified and controlled unclassified geospatial products, topographical, nautical, and aeronautical data, products and services. Both side are working on it.

Harsh V. Pant, professor of International Relations at King's College, London, highlighted the importance of India to US strategic planning by saying: "India is key to the US' ability to create a stable balance of power in the larger Indo-Pacific and at a time of resource constraints, it needs partners like India to shore up its sagging credibility in the region in face of Chinese onslaught."[24] The USA, as part of its foreign policy to counter China, wants to make India one of its major defence partners for which it is in talks with Indian representatives to sell highly technologically advanced equipment including predator drones. Since 2008, India has purchased the P 8I maritime patrol aircraft, the C-17 and C-130 aircraft, the Chinook and Apache helicopters, M777 Howitzers, and more recently the National Advanced Surface to Air Missile System (NASAMS), version II, multi-layered air defence shield and 24 MH-60 Sikorsky Romeo multi-role helicopters.

US Disengagement in Afghanistan

As USA tries to disengage in Afghanistan, after a failed policy in the region, it literally hands over the reins to the Pakistan-controlled Taliban, who have intensified attacks immediately after announcement of the likely withdrawal. This included the March 24, 2020, attack by heavily-armed terrorists on a group of nearly 200 Sikh and Hindu worshippers in the gurdwara at Shor Bazaar in Kabul city in which 19 people died.[25] Though this attack was claimed by the ISIS, the July 2018 attack on the Sikh and Hindu delegation in Jalalabad, in which 19 were killed and 21 injured, was claimed by the Taliban. With American withdrawal and Pakistan support, the Taliban and ISIS could try to penetrate India further. China is relying on Pakistan for support to get a foothold in the Middle East. The two are sizable military powers and pose a significant threat to India. As China bullies its way through, India needs to form alliances to strengthen its position, especially in the IOR.

MILITARY ASSETS AND APPROACH INDO-PACIFIC

United States Indo-Pacific Command (INDOPACOM)

INDOPACOM[26] is a unified combatant command of the U.S. Armed Forces for the Indo-Pacific region. It is the oldest and largest of the unified combatant commands, covering 260,000,000 sq. km, or roughly 52 per cent of the Earth's surface, stretching from the U.S. West Coast to the west coast of India, and from the Arctic to the Antarctic. Its components include the U.S. Army Pacific, Marine Forces Pacific, the U.S. Pacific Fleet, Pacific Air Forces (PACAF), U.S. Special Forces Command Pacific, and U.S. Forces in Japan and Korea. The INDOPACOM headquarters is located at Hawaii. Originally, U.S. PACOM, the command was renamed as U.S. Indo-Pacific Command on 30 May 2018, in recognition of the greater emphasis on South Asia and India. Its area of responsibility stretches from San Diego to Diego Garcia, from the Pacific to the Eastern Indian Ocean (68 degrees east), including India. Traditionally the Commander has always been from the U.S. Navy.

PACAF is the air component command. Over the past 65+ years,

PACAF had engaged in combat during the Korean and Vietnam wars and Operations Desert Storm, Southern Watch, Northern Watch, Enduring Freedom and Iraqi Freedom. The mission of 4,500 strong PACAF is to provide ready air and space power to promote U.S. interests in the Indo-Pacific region. PACAF comprises three numbered Air Forces, nine main bases and nearly 375 aircraft. A much larger American air component is with the USN's U.S. Pacific Fleet which consists of approximately 200 ships (including five aircraft carrier strike groups), nearly 1,100 aircraft, and more than 130,000 sailors.

PACAF Operating Assets

PACAF has the Fifth Air Force at Yokoto airbase in Tokyo, Japan. It has two major airbases at Kadena and Misawa, Japan, with a cross-section of combat air assets including fighters, FRA, AEW&C, EW and Recce. The airbase at Yokota in Japan has Special Ops aircraft. The Seventh Air Force headquarters is at Osan airbase in South Korea. Osan and Kunsan have fighter wings with significant combat air assets. The second airbase in Korea is at Kunsan. The Eleventh Air Force is headquartered at Elemndorf airbase in Alaska. It has air assets in two major airbases in Alaska. They also operate from a joint base at Pearl Harbour-Hickam. PACAF also has an Air Force Reserve Command. There is the Anderson airbase at Guam.

US Navy - Aircraft Carrier CVN 78 - USS Gerald R. Ford.
Image Source: seaforces.org

U.S. Pacific Deterrence Initiative[27] (PDI)

The US Indo-Pacific Command's $20 billion wish list (over 6 years) to deter China has been presented to the U.S. Congress. This includes a $ 1.6

billion defensive ring around Guam, millions in new military funding for partner nations, and a billion dollars for increased stockpiles of long-range weapons. The new strategy to "Regain the Advantage is designed to persuade potential adversaries that any pre-emptive military action will be extremely costly and likely to fail by projecting credible combat power at the time of crisis, and will provide the President and Secretary of Defence with several flexible deterrent options to include full operation plan execution, if necessary." The suggestion is that a Pacific Deterrence Initiative, or PDI, would be focused on dealing with China in the INDOPACOM region. It's not just a budget exercise, but is "a broader strategic opportunity to message the U.S. commitment to the Asian allies and partners as well as to Beijing who have grown confident of their military capabilities. PDI can ensure U.S. forces have what they need to deter Chinese aggression and maintain a free and open Indo-Pacific.

The PDI is likely to be more focused on Navy and Air Force needs. The proposed funding will be for five broad categories. Joint force lethality; force design and posture; strengthening of allies and partners; exercises, experimentation and innovation; and logistics and security enablers. It would mean fielding an integrated Joint Force with precision-strike networks, particularly land-based anti-ship and anti-air capabilities along the First Island Chain; integrated air missile defence in the Second Island Chain; and an enhanced force posture that provides for dispersal, the ability to preserve regional stability and, if needed, sustain combat operations. There is a proposal for a 360-degree persistent and integrated air defence capability in Guam.

The proposal includes integration of long-range precision fires such as the Navy's Maritime Strike Tomahawk and the Air Force's JASSM-ER weapon; a high-frequency radar system based in Palau to detect air and surface targets; a homeland defence radar in Hawaii to detect ballistic, cruise and hypersonic threats; and a space-based persistent radar system for tracking global threats. The combatant command "requires highly survivable, precision-strike networks along the First Island Chain, featuring increased quantities of allied ground-based weapons", Adm. Phil Davidson, head of INDOPACOM, wrote.

Infrastructure investments are needed to spread the U.S. military around the region, breaking the longstanding network of large, centralized bases now seen as easy targets for China's long-distance capabilities. America's longstanding advantage in the Pacific relies on a bedrock of alliances and partnerships in the region. Setting up a Mission Partner Environment, envisages the use of "cloud-based technologies, integrated systems, and secure access controls to provide assured command, control, and communications (C3)," and the creation of three fusion centres to work with allies on specific tasks, including a counterterrorism cell already in the works with Singapore and others. The need is for dispersal locations, airfield battle-damage, repair capabilities, and infrastructure for C4I, munitions generation, mobility processing, and fuel storage.

U.S. Other Regional Assets

The USA will like to continue to be an Indo-Pacific power, retaining territorial possessions across the Pacific as well as bases in the Indian Ocean. Guam sits in the "second island chain" which runs from Japan's Bonin Island, and is considered the US "tip of the spear". Its deep water facilities enable it to handle aircraft carriers, and it has a long airstrip at Andersen airbase, able to house heavy strategic B-52 bombers. Guam was the host site for the trilateral *Malabar* exercises in June 2018 between the USA, India and Japan, with the presence of the USS *Ronald Reagan*, the lead US nuclear-powered aircraft carrier.[28] The USA has a significant presence in Japan, with a Carrier Group at Yokusaka and Ryukyu chain of islands, just off Taiwan. Taiwan and Japan make a great strategic combination immediately next to China. US basing facilities at strategically important Palawan in the Philippines facing the South China Sea were re-established in 2016. The U.S. has berthing facilities at Da Nang in Vietnam, not too far from Chinese Hainan Island. The USS *Carl Vinson* carrier strike group made a historic visit to Da Nang in March 2018, as a powerful message to Beijing. The Cope West exercises between the Indonesian and US air forces have run annually since 2012, in Indonesia or at Tinian in the Western Pacific north of Guam. Indonesia is described as "a geographic and diplomatic fulcrum for the Indo-Pacific region."[29]

The USA has defence links with Singapore[30] established under the 2005 Strategic Framework Agreement and further strengthened under the 2015 Enhanced Defence Cooperation Agreement have resulted in an ongoing Logistic Group West Pacific stationed there by the USA. Special aircraft carrier berthing facilities, ongoing deployment of littoral combat warships and regular aircraft deployments at Singapore enable further U.S. projection into the Eastern Indian Ocean and South China Sea.

Similar Indo-Pacific projection is enabled with the Marine Rotational

Force agreed with Australia at Darwin in November 2011.[31] In turn, the UK atoll of Diego Garcia has been the site of a significant U.S. base since 1977, embedding US maritime power in the middle of the Indian Ocean, and anchoring America's future presence in the Indo-Pacific. The USA is a member of various organizations in the Pacific and Indian Oceans; including the East Asia Summit (EAS), the ASEAN Regional Forum (ARF), and the ASEAN Defence Ministers Meeting (ADMM).

Japan Air Self-Defence Force

The Japan Air Self-Defence Force, also referred to as the Japanese Air Force, carries out combat air patrols around Japan, and maintains a network of ground and air early-warning radars. It is estimated to have around 50,000 personnel and operates around 750 aircraft, approximately 375 of them fighters. The force will be renamed the Japan Aerospace Self-Defence Force in recognition of the increasing importance of the space domain. Their fighter aircraft include the Mitsubishi F-2 (based on F-16), F-15 Eagle, and F-35 Lightning II. The Boeing E-767 and E-2 Hawkeye are AEW&C aircraft. They also have about six FRA and a few transport aircraft. The helicopters include CH-47 Chinook and Sikorsky UH-60.

F-35A Lightning II.
Picture Source: Wikipedia

The Royal Australian Air Force

The Royal Australian Air Force provides support across a spectrum of operations such as air superiority, precision strikes, intelligence, surveillance and reconnaissance, air mobility and space surveillance. It has 259 aircraft, of which 110 are combat ones. Combat aircraft include

EA-18, F/A-18A/B Hornet, F/A-18 Super Hornet and F-35A Lightning II. Other aircraft include AP-3C Orion, P8A Poseidon, C 130J Hercules, C-17A Globemaster III and KC-30A MRTT, among others. Many more F-35A and Super Hornets will be inducted.

Vietnam People's Air Force

The main mission of the Vietnam People's Air Force is the defence of Vietnamese airspace and the provision of air cover for operations of the People's Army of Vietnam. It has nearly 300 aircraft, mostly of Russian origin, and has 5,000 personnel. Its aircraft include Sukhoi Su-22, Su-27, and Su-30. The transport fleet includes nearly 30 Ukrainian An-26. They have sizeable numbers of Mi-8/Mi-17 helicopters and also Mi-24 attack helicopters.

Air Balance Indo-Pacific

The air forces of the QUAD group of countries and also Vietnam have sizeable numbers and presence. Most air forces have state-of-the art air assets. Each one is individually capable of acting as a deterrent. China is conscious that the rationale of the group of QUAD nations is encirclement and containment of China and that the grouping has sizeable military strength. China has the advantage of home ground and increasing new generation air assets, but still lags in some core technologies like the aircraft engine. It still has a very small aircraft carrier force to be of any significance in the Indo-Pacific region. Undoubtedly, they will catch up in a decade or so.

RIMPAC

RIMPAC (Rim of the Pacific) Exercise, is the world's largest international maritime warfare exercise held biennially (even-numbered years) from Hawaii. Hosted by the USN's Indo-Pacific Command. Australia, Canada, New Zealand, the UK and the U.S. participate. Other regular participants are Chile, Colombia, France, Indonesia, Japan, Malaysia, the Netherlands, Peru, Singapore, South Korea, and Thailand. The New Zealand Navy is also frequently involved. Several observer nations are usually invited, including China, Ecuador, India, Mexico, the Philippines, and Russia.

National Security Strategy of the USA

The policy released in December 2017 contained a specific section on

"The Indo-Pacific".[32] It warned that "geopolitical competition between free and repressive visions of world order is taking place in the Indo-Pacific region, which stretches from the west coast of India to the western shores of the USA" in which "China seeks to displace the United States in the Indo-Pacific region", but for which "the United States must marshal the will and capabilities to compete and prevent unfavourable shifts in the Indo-Pacific". It was argued that this was to be achieved through forward deployment of US forces complemented by quadrilateral arrangements with Australia, India and Japan, together with other bilateral arrangements with countries like Indonesia, Singapore and Vietnam. Similarly, the *2018 National Defence Strategy*[33] talked of strengthening alliances and attracting new partners.

Japan, India's Key Strategic Partner[34]

Trump has come down harshly on China's mercantilist trade policies by imposing destabilising tariffs on Chinese exports. Russia's global partnership with China is just to balance American power, and does not materially affect Russia's close relations with India. Amidst all this volatility in global power equations, India's most important partner across its Indian Ocean neighbourhood currently is Japan. Both India and Japan have been challenged by Chinese territorial claims, actions and ambitions, together with China's quest to become a hegemonic power in Asia. India and Japan also closely cooperate on their relationship with China, including on measures to see that tensions with China do not get out of control. Prime Minister Modi and President Xi Jinping have had "one to one" meetings at Wuhan and Mamallapuram to "issue strategic guidance to their respective militaries to strengthen communication to build trust and understanding, to implement various confidence-building measures, which have already been agreed upon, by the two sides". Yet Chinese military has shown aggressive posturing and created conflict like situation in Galwan region of Ladakh. This has created trust deficit between the two sides. Japanese Prime Minister Abe and President Xi Jin Ping focused on avoiding actions that could escalate tensions across disputed maritime boundaries in the East China Sea. There has been a measure of congruence in the approach of India and Japan to maintain peace and tranquillity across their respective land and maritime boundaries with China. India is working in close cooperation with Japan, on economic development projects across the Indian Ocean region to ensure that countries in the region do not become overly dependent on China. The Indo-Japanese partnership can also balance the vast resources that China is committing primarily for infrastructure development in India's South Asian and Indian Ocean neighbourhood.

CPTPP and RCEP

There are also strategic economic alignments in the Indo-Pacific. The two largest regional trade agreements today are the Regional Comprehensive Economic Partnership (RCEP) and the Comprehensive and Progressive Agreement for Trans-Pacific Partnership (CPTPP). ASEAN is vital to both. All ten members are part of RCEP, while four, (Singapore, Malaysia, Vietnam and Brunei) are also part of the Canada-led CPTPP. India which is eligible for RCEP membership has opted out currently.

OPTIONS AND CHALLENGES INDIA

Balancing China: Ground Realities[35]

China's neighbours "are certain to fear its rise" and "will do whatever they can to prevent it from achieving regional hegemony," including joining "an American-led balancing coalition to check China's rise." Most of China's neighbours and peers are struggling to remain as diplomatically and economically engaged as possible with the rising Asian power while taking the minimum steps necessary to preserve their security and sovereignty. Many, are averse to treaty alliances, multilateral military exercises, and joint Freedom of Navigation Operations and naval patrols. Even the members of the "Quad" have reservations about a Cold War-style containment strategy.

Most of China's neighbours prefer a balancing approach. Regional defence collaboration has been limited to joint military exercises; security-focused bilateral, trilateral, and multilateral dialogues; joint vision statements; and military inter-operability agreements. China's actions have pushed Quad members together because of growing concerns about its challenges to the rules-based order, as in SCS. Yet, for most Indo-Pacific countries, trade and investment ties with China have grown exponentially in recent years. Even among the Quad, diplomatic and economic cooperation with China remains remarkably robust; Beijing is the largest trading partner of all four members. Some of China's immediate neighbours, like Cambodia, Laos, and Pakistan, have forgone balancing altogether, and are actually closing up. ASEAN states, meanwhile, have steered away from hard balancing measures toward more neutral, non-provocative endeavours. They're seeking to defend their sovereignty and autonomy by strengthening international laws, norms, and institutions, which realists have traditionally dismissed as inconsequential to restraining state behaviour and impacting decisions about war and peace. Severing ties with China is "politically, economically, and practically unthinkable" in a globalised world, leaving China free to exploit interdependence to increase power. Most are just vocally insisting on rule-based order and a free and open Indo-Pacific including freedom of navigation, peaceful dispute settlement, and support for international laws that constrain China's most aggressive impulses and its capacity for misbehaviour.

While Beijing freely disregards laws, norms, and international opinion, it spared no effort trying to prevent ASEAN from forming a consensus in opposition to its South China Sea activities. Most Chinese neighbours, other than Taiwan, don't fear a kinetic action or Chinese-funded insurgencies; they believe China poses a threat to their interests, autonomy, and indulges in grey-zone coercion designed to induce submission. Most Southeast Asian countries will continue taking modest steps to strengthen their military capabilities and forge new external security partnerships. Some, like Vietnam, may pursue more independent foreign policies and more vigorous balancing initiatives, drawing closer to the Quad. Indonesia has shown some support with the unveiling of an Indo-Pacific cooperation concept. Some ASEAN member-states are concerned that Quad will leave ASEAN out in the cold. The failure to release a joint communiqué on the South China Sea at the ASEAN Ministers' Meeting in 2012 was a notable example of limitations. Within the Quad, all four countries have been advancing defence collaboration at the bilateral and trilateral levels in much quicker but quieter strides. Growing concerns about China's expansive Belt and Road Initiative will be an issue in coming years. There is greater scrutiny applied to Chinese investments in sensitive infrastructure. The Trump Administration has opened several new fronts against China on economic protectionism, cyber-security, and human rights violations.

China's Response to Indo-Pacific Concept

Although the Chinese government has not openly discussed the United States' Indo-Pacific strategy in China, the Chinese academic community has had a heated debate about the nature of the concept, the possible impact on China and the region, and the future of U.S.-China relations.[36] According to some Chinese scholars, the Indo-Pacific strategy is a preliminary idea for the USA to connect the Indian Ocean and the Pacific region, to constrain China's rise from a geopolitical perspective, and to safeguard its own leadership and interests in the region. Many Chinese scholars believe that the concept has not matured. Wang Xiaowen, a scholar at Beijing Language and Culture University, wrote that the Indo-Pacific strategy is essentially an extension and deepening of the "rebalance," with the aim of strategically linking the Indian and Pacific oceans. The Chinese foreign minister reportedly has stated that the idea would "dissipate like foam". Yet China has made significant overtures in the eastern Indian Ocean due to the economic value of the sea lines of communication passing through it. China has not only invested in several ports but has also increased its naval forays in the eastern Indian Ocean.

Challenges and Options for India

Rajesh Rajagopalan, a professor of International Politics at Jawaharlal Nehru University in New Delhi, looks at India's strategic choices.[37] China is a direct military threat to India, particularly in the light of the border disputes between the two countries. Though India has considerable military power, China's forces are already stronger and better-funded; Beijing's outsized wealth will likely allow it to outspend New Delhi in the foreseeable future. Beijing's influence in both established international organizations like the United Nations and in the new institutions China is setting up, such as the Asian Infrastructure Investment Bank, gives Beijing opportunities to hamper Indian interests and goals in multilateral forums, especially when it comes to reforming these institutions and giving India a greater voice in global affairs. China's alignment with Pakistan and deepening relations with other South Asian countries represents a significant challenge to India's position in the region, which otherwise New Delhi had dominated for decades. Beijing's ability to provide financial assistance and balance against New Delhi may tempt India's smaller neighbours to play one power against the other, undermining India in its own backyard. China's economic power allows Beijing to spread its influence around the world, which could be used to India's detriment.

India's options have to be based on China's growing military muscle, its power in international institutions, including the UNSC and NSG. China is establishing international institutions like the Asian Infrastructure Investment Bank and shaping other multilateral organizations to promote Chinese interests, such as the BRICS and the SCO. China's willingness to play the role of an external balancer against India in South Asia is a serious challenge and can be seen as a military threat in support of Pakistan. Some smaller states use China to counter India's natural domination of the region. China's great economic power allows it to spread its influence around the world, sometimes to India's detriment.

India must decide how to secure its interests in this unbalanced environment by choosing among six potential strategic options: staying unaligned, hedging, building indigenous military power, forming regional partnerships, aligning with China, or aligning with the USA. A strategy of nonalignment, hedging, or alignment with China would in all likelihood not serve India's interests because China's power, geographical proximity, and policies already represent a clear danger to India's security and global interests. A closer alignment with Washington is likely to represent India's best chance to counter China, while efforts to foster regional partnerships and cultivate domestic military capabilities, although insufficient by themselves, could play a complementary role. The USA is the only power that is stronger than China. Further, New Delhi and Washington share a common interest in balancing Beijing. Among India's other strategic options, efforts to build indigenous military power and forge regional partnerships are necessary and complementary means of countering China, but are by themselves insufficient, because China is already wealthier and stronger than India or any combination of other Asian powers. India has to make a choice because strategic capabilities have long lead times and cannot be built up quickly. India cannot find a suitable strategic partner after a crisis has already developed. China's geographic proximity to India coupled with its military strength severely impinges on Indian security and constrains the possibility of New Delhi pursuing nonalignment. Considering that the balance of power is heavily tilted in favour of China, India requires relationships that go beyond being arms suppliers. India needs strong partners who can not only coordinate with India to balance China's military power but also counter its political and economic clout in multilateral institutions. India's suspicion of alliances and its desire for strategic autonomy is understandable. Can India afford a wait and watch policy? Unlikely.

In the meantime, India has to build up its military strength for its own security, a form of insurance that cannot be ignored. Building sufficient military capabilities is necessary to act as a sufficient deterrent for Beijing to not open a second front in case of war with Pakistan. India also needs strong partners who can help balance against China and possibly help India enhance its own capabilities. India does not currently have sufficient military capability to counter China on its own. Beijing continues to outspend New Delhi on defence manifold. China clearly has a technological edge over India. It does have multiple threats, especially from the USA and Japan. Considering that India's military objective against China is defence and deterrence and not offensive in nature, it reduces the military burden. Ultimately, India has little choice but to enhance its military capabilities to the extent that it can.

Regional balancing is a strategy India must pursue to align with other Asian countries such as Australia, Japan, the Philippines, Singapore, and Vietnam. Later, Indonesia and Malaysia could also be incorporated. All these countries are also concerned about China's rise and aggressiveness. India's inability to improve transportation infrastructure in its east is a serious problem. A regional balancing approach allows India to balance China without direct alignment with the USA. Unlike in the case of alignment with a major power, India would be the more powerful partner in regional relationships. But China is much too strong already for regional states to balance against it. For India to rely solely or predominantly on regional balancing to ensure its security, then, would be exceedingly difficult and risky.

So, ultimately, while India pursues regional balancing and military strength, aligning more closely with the USA without any formal treaty alliance could be the choice. This exercise will not be easy because of years of suspicion and also India's tacit alignment with the Soviet Union and Russia. India's close relationship with the USA is just a decade old. India needed the support of the USA for the NSG waiver, and high technology military equipment. Any alignment with it will be complex and mostly for the need to tackle the challenge of China's rise. The most important benefit of deepening a partnership with the USA would help India balance China, because no other country can help it to do so. The USA is powerful, and has a self-interest in partnering with India. It has a global network of alliances and partnerships. Russia, India's traditional strategic partner, does not have such strength or global influence. Yes, there is a fear that China may overtake USA as the leading global power. The USA will have to balance China not only because China seeks hegemony in Asia but also because none of the other Asian powers are likely to be strong enough to balance it. As China becomes stronger, Indian and U.S. interests in balancing China will only grow. As the U.S.-China power differential narrows, the American imperative for balancing China is likely to increase, as is Washington's need for allies. The U.S. capacity to advance Indian interests in multilateral forums is of interest to India. India will need U.S. support in its pursuit of a permanent seat on the UNSC and possibly on other global governance issues. The USA remains the world's leader in high-tech research and development, and this is particularly true when it comes to advanced weapons technology. India needs such technology. There seems an evolving political acceptance of such a partnership. Any partnership would have to balance India's relationships with others, especially Russia. Much to Russia's dislike, currently the USA has pushed it into Chinese arms. China's balancing and containment strategy against India would not cease even if India did not partner with the USA. Meanwhile, India and USA have different geo-political positions on many regional disputes which would have to be put aside. The USA has partnered or jettisoned many allies in the past, including Russia, China, Iran and Pakistan, and this also raises some doubts.

Vivek Mishra writes about India's nuanced Indo-Pacific Strategy.[38] "Although India has created a robust strategic arc from the Persian Gulf and the Asia-Pacific, its Indo-Pacific strategy is more nuanced, involving careful balancing between multiple stakeholders. For instance, though India has promised the United States and its ilk that it will help to shape a regional strategy in line with their multilateral Indo-Pacific vision, its multifaceted relationships with China and Russia demand that New Delhi also create latitude for Beijing and Moscow within this vision. Addressing the Shangri La Dialogue in June 2018, Prime Minister Narendra Modi confessed that "no other relationship of India has as many layers as our relations with China." The acknowledgement is reflected in India's Indo-Pacific strategy, which avoids any direct confrontational stance vis-à-vis China while partnering with the USA to create a regional structure that is inconsistent with Chinese expectations. While India has readily attended the Quad's meetings, it has also restored its maritime dialogue with China. In navigating the competing interests of the USA, China, Russia, and other players in the region, India risks being deemed unreliable by its partners. New Delhi should attempt to manage the expectations of its partners such that they are cognizant of India's apprehensions and limitations, while the USA and its allies should accommodate India where possible.

REFERENCES

1. ANI, Indo-Pacific region is our lifeline, says PM Modi, *Business Standard*, June 8, 2019, https://www.business-standard.com/article/news-ani/indo-pacific-region-is-our-lifeline-says-pm-modi-119060800814_1.html
2. New Delhi Times Bureau, Renaming US Pacific Command as Indo-Pacific: Implications for India, July 09, 2018, https://www.newdelhitimes.com/renaming-us-pacific-command-as-indo-pacific-implications-for-india/
3. Ibid.
4. Gurpreet S. Khurana, 'Security of Sea Lines: Prospects for India-Japan Cooperation', *Strategic Analysis* (IDSA/ Routledge), Vol. 31 (1), January 2007, pp.139-153.
5. Australian Government, Department of Defence, Defence White Paper 2013, p. 7, https://www.defence.gov.au/Whitepaper/2013/docs/WP_2013_web.pdf
6. White House Foreign Policy Briefing, United States and India: Prosperity Through Partnership, June 26, 2017, https://www.whitehouse.gov/briefings-statements/united-states-india-prosperity-partnership/
7. 2017 Foreign Policy White Paper, Australian Government, Pages 28 and 40, https://www.globalsecurity.org/military/library/report/2017/australia_2017_ foreign_policy_white_paper.pdf
8. Authors: Jonathan Woetzel, Shanghai; Jeongmin Seong, Shanghai; Nick Leung, Hong Kong; Joe Ngai, Hong Kong; James Manyika, San Francisco; Anu Madgavkar, Mumbai; Susan Lund, Washington, DC; and Andrey Mironenko, Sydney. China and the world: Inside the dynamics of a changing relationship (In Brief) McKinsey Global Institute. July 2019, https://www.mckinsey.com/~/media/mckinsey/featured %20 insights/china/china%20and%20the%20world%20inside%20the%20 dynamics%20 of%20a%20changing%20relationship/mgi-china-and-the-world-full-report-june-2019-vf.ashx
9. Bloomberg, Australia warns of Asia power shift to China if US withdraws from region, *Financial Express*, November 23, 2017, https://www.financialexpress.com/world-news/australia-warns-of-asia-power-shift-to-china-if-us-withdraws-from-region/944865/
10. A FREE AND OPEN INDO-PACIFIC Advancing a Shared Vision, A U.S. State Department Document, November 04, 2019, https://www.state.gov/wp-content/uploads/2019/11/Free-and-Open-Indo-Pacific-4Nov2019.pdf
11. Nitin Pai, India and the Indo-Pacific Balance of Power, *Indian National Interest*, August 01, 2017, https://nationalinterest.in/india-and-the-indo-pacific-balance-of-power-1814c5a5f6a1
12. Jacob Goldberg, What Is India's Look East Policy? ThoughtCo.com. Updated November 12, 2019, https://www.thoughtco.com/look-east-policy-of-india-1435050
13. Strachan, Anna Louise, Harnit Kaur Kang, Tuli Sinha, and Rajiv Sikri. Report. Institute of Peace and Conflict Studies, 2009. Accessed March 31, 2020. www.jstor.org/stable/resrep09289.
14. Nitin Pai, India and the Indo-Pacific Balance of Power, *Indian National Interest*, August 01, 2017, https://nationalinterest.in/india-and-the-indo-pacific-balance-of-power-1814c5a5f6a1

15. Rajesh Rajagopalan, Evasive balancing: India's unviable Indo-Pacific strategy, *International Affairs,* Volume 96, Issue 1, January 2020, Pages 75–93, https://academic.oup.com/ia/article/96/1/75/5697517
16. Ibid.
17. Nitin Pai, India and the Indo-Pacific Balance of Power, *Indian National Interest,* August 01, 2017, https://nationalinterest.in/india-and-the-indo-pacific-balance-of-power-1814c5a5f6a1
18. Dipanjan Roy Chaudhury, India's geopolitical status goes up after Doklam standoff ends, *Economic Times,* July 13, 2018, https://economictimes.indiatimes.com/news/defence/indias-geopolitical-status-goes-up-after-doklam-standoff-ends/articleshow/60282585.cms?from=mdr
19. Bharath Gopalaswamy, Aditya Ramachandran, The Shifting Balance of Power in the Indo-Pacific, chinausfocau.com, December 19, 2017, https://www.chinausfocus.com/peace-security/the-shifting-balance-of-power-in-the-indo-pacific
20. James McBride and Andrew Chatzky, What is the Trans-Pacific Partnership 9TPP)? Council on Foreign Relations, Updated January 04, 2019, https://www.cfr.org/backgrounder/what-trans-pacific-partnership-tpp
21. David Scott, The Indo-Pacific in US Strategy: Responding to Power Shifts, *Rising Powers Quarterly,* Volume 3, Issue 2, August 2018, p. 20, https://pdfs.semanticscholar.org/8891/ee0e0197b4bccebcae4e5684815124ba2fd5.pdf?_ga=2.2697648.1605024220.1585743187-1699218414.1583153731
22. Dipanjan Roy Chaudhury, India's fine balancing act with Quad and BRICS meet in New York, *Economic Times,* September 28, 2019.
23. Amb. Sujan R. Chinoy, Indo-US Defence Partnership: Future Prospects, Manohar Parrikar Institute for Defence Studies and Analyses, New Delhi, June 26, 2019, https://idsa.in/idsacomments/indo-us-defence-partnership-srchinoy-260620
24. India–United States relations, Wikipedia, https://en.wikipedia.org/wiki/India%E2%80%93United_States_relations
25. Kabul gurdwara attack turns spotlight on CAA, *The Tribune,* March 30, 2020, https://www.tribuneindia.com/news/comment/kabul-gurdwara-attack-turns-spotlight-on-caa-62986
26. United States Indo-Pacific Command, wikipedia, https://en.wikipedia.org/wiki/United_States_Indo-Pacific_Command
27. Aaron Mehta, Inside US Indo-Pacific Command's $20 billion wish list to deter China —and why Congress may approve it, defensenews.com, April 02, 2020, https://www.defensenews.com/global/asia-pacific/2020/04/02/inside-us-indo-pacific-commands-20-billion-wish-list-to-deter-china-and-why-congress-may-approve-it/?utm_source=Sailthru&utm_medium=email&utm_campaign=Breaking%20News%2004.02.20&utm_term=Editorial%20-%20Breaking%20News
28. David Scott, The Indo-Pacific in US Strategy: Responding to Power Shifts, *Rising Powers Quarterly,* Volume 3, Issue 2, August 2018, Pages 21-29, https://pdfs.semanticscholar.org/8891/ee0e0197b4bccebcae4e5684815124ba2fd5.pdf?_ ga=2.2697648.1605024220.1585743187-1699218414.1583153731
29. Ibid.
30. Ibid.
31. Ibid.
32. Ibid.
33. Summary of National defence Strategy 2018, US Department of Defence, https://

dod.defense.gov/Portals/1/Documents/pubs/2018-National-Defense-Strategy-Summary.pdf

34. G. Parthasarathy Emerging balance of power in Asia, *The Hindu* Business Line, November 28, 2018, https://www.thehindubusinessline.com/opinion/columns/g-parthasarathy/emerging-balance-of-power-in-asia/article25616068.ece
35. Jeff Smith, China's Rise and Balancing in the Indo-Pacific, The Heritage Foundation, January 09, 2019, https://www.heritage.org/asia/commentary/chinas-rise-and-under-balancing-the-indo-pacific-putting-realist-theory-the-test
36. Dingding Chen, Chinese scholars see the free and open Indo-Pacific strategy as containment, but aren't convinced it'll stick, *The Diplomat*, April 27, 2018, https://thediplomat.com/2018/05/what-china-thinks-of-the-indo-pacific-strategy/
37. Rajesh Rajagopalan, India's Strategic Choices: China and the Balance of Power in Asia, Carnegie India, September 14, 2017, https://carnegieindia.org/2017/09/14/india-s-strategic-choices-china-and-balance-of-power-in-asia-pub-73108
38. Vivek Mishra, India's nuanced Indo-Pacific strategy, *South Asian Voices*, November 1, 2018, https://southasianvoices.org/india-indo-pacific-strategy/

10

Air War Across the Himalayas

"I will come again and conquer you because as a mountain you cannot grow, but as a human, I can."

—**Sir Edmund Hillary,** the first man to climb the Everest

"If Ganges is the mother, Himalayas is the father. One nurtures and nourishes, the other provides and protects."

—**Vinita Kinra**

The Himalayas comprise the most dominating geographical feature of India. No other mountain range anywhere in world has affected the life of the people and shaped the destiny of a nation as the Himalayas have in respect of India. As per ancient texts, the Himalayas are the body and soul of India. In a very special measure, the Himalayas have been protecting India from outside invaders since the early times, thus serving as a defence barrier. Tibet was the buffer between the two Asian giants until it was annexed by China in 1951. Later, the Chinese aggression on India in October 1962 reduced the defence significance of the Himalayas to a considerable extent. Yet, despite advances in modern warfare technology, the defence significance of the Himalayas cannot be ignored altogether. The long border remains highly militarised.

"The Hump" was the nickname Allied pilots gave the airlift operation that crossed the Himalayas, the "Roof of the World", from India into China during WW II.[1] It was the US Army Air Force's most dangerous airlift route. More than 1,000 men and 600 planes were lost over the 530-mile stretch of rugged terrain. It was dubbed the "Skyway to Hell" and the "Aluminium Trail" because of the number of planes that didn't make it.

Friendship with the two Himalayan countries, Bhutan and Nepal, have been cornerstones of Indian foreign policy. However, the bigger challenge is overlapping Indian and Chinese space. Ironically, both ancient nations are locked in a boundary dispute in the Himalayas. Prime Minister Modi, in his speech in the Nepali Constituent Assembly, said the Himalayas hold the keystone for Asian culture, environmental, political, and regional security.

China's Neighbour Troubles

China with the world's largest standing army and second largest defence budget has traditionally been a hegemonic power and is still pushing 23 boundary disputes with its 14 neighbours. Territorial disputes with other countries include Taiwan; Yalu river enclaves with North Korea; Aksai Chin and Arunachal Pradesh with India; Four pockets of land in Bhutan; the Japanese Senkaki Islands; Paracel Islands with Vietnam; Scarborough Shoal with the Philippines; and a complex dispute over Spratly Islands with four countries. The India-China Hotline between Indian Army DGMO and Western Theatre Command Commander is likely to be operational soon.

Chinese Muscle Flexing

The Doklam incident on the China-Bhutan-India border had once again indicated China's muscle-flexing unilateral approach to solving border disputes. The more recent Galwan incident in Ladakh is another indicator of unilateralism. Earlier, it had usurped nearly 90 per cent of the SCS by creating seven man-made islands and now claims nearly 3 million square kilometres as its exclusive economic zone (EEZ) with utter disregard to the ruling of the International Arbitration Tribunal which saw meek world response.[2] China's SCS stand has serious security and economic implications. Belligerent China's containment or building a deterrent against it is important for all. Meanwhile, India has come a long way and is today the fastest-growing large economy and globally among the first four countries in most economic and military indicators. India has a well-placed nuclear triad and has an Anti-Ballistic Missile (ABM) system. The Indian Armed Forces are significantly strong and in many sectors are better

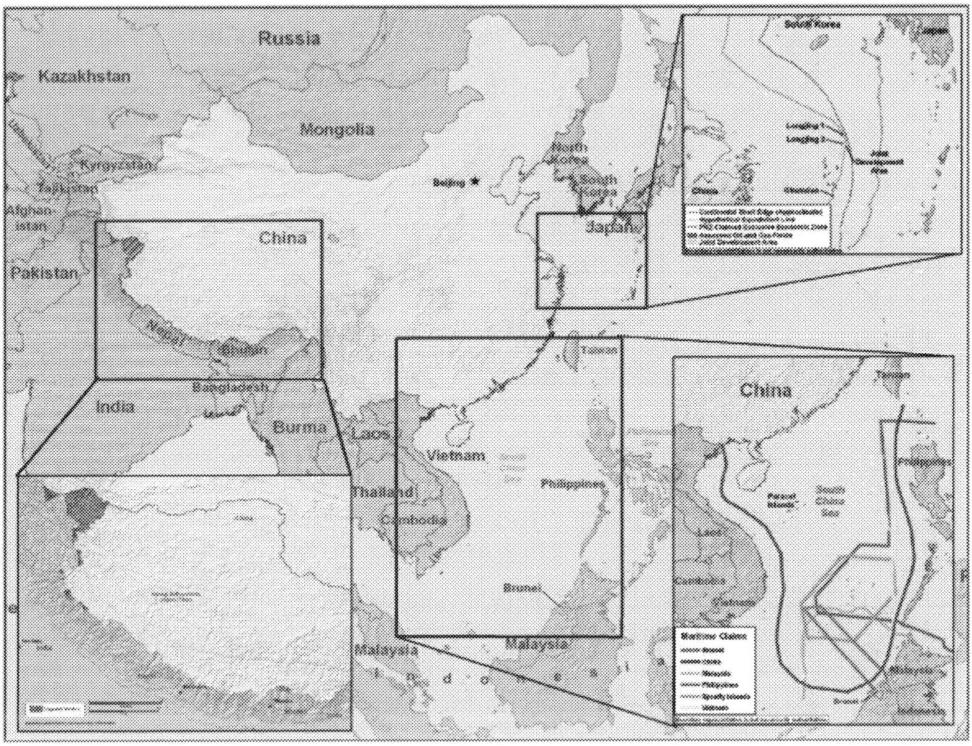

China's Boundary Disputes
Image Source: Australian Defence Association

placed than the Chinese across the Himalayan frontier. While China is hailed as a potential new superpower, some warn that economic bubble and demographic imbalances could slow or even halt China's growth as the century progresses.

String of Pearls

For over a decade, China has been investing in the Indian Ocean region and using geopolitical influence through economic inducement, military supplies, airfields and port construction. Popularly known as the 'String of Pearls', India sees it as the encirclement of itself.[3] Chinese investments in Gwadar port in disputed Balochistan in Pakistan, and the BRI road-rail corridor to connect through disputed Pakistan occupied Kashmir has been questioned by India. Hambantota port in Sri Lanka, Chittagong and Sonadia in Bangladesh, a port in Sudan on the Red Sea, a port in Somalia and now its first overseas military base in Djibouti are part of dominating the Indian Ocean region. China has for long been building roads and airfields in Myanmar and developed a deep-water port and electronic intelligence base at Coco Islands, just 18 km from India's Andaman Islands.

Indian Strategic Partnership Initiatives

China became the world's largest economy in purchasing-power-parity (PPP) terms in 2014.[4] In nominal terms, China's GDP was 66 per cent of the USA in 2018, making it the second-largest economy in the world. On the MGI Connectedness Index that ranks participation by flows of goods, services, finance, people, and data, China was the ninth most-connected country in the world in 2017. In 2018, it accounted for about 16 per cent of the world's GDP. The USA is the world's most technologically powerful economy and the largest by nominal GDP and net wealth and the second-largest by PPP. Americans hold nearly 40 per cent of the total wealth, accounting for nearly a quarter of the global GDP and a third of global military spending. The USA's 2019 defence budget at $ 693 billion compares with US$ 177 billion for China, and $ 61 billion for India. Since early 2000, the USA has been trying to befriend India to contain China. India has been engaged in an aggressive round of international diplomacy forging one-to-one equations and strategic partnerships with the USA, Russia, Japan, France, Australia, and Israel among many others.

China Balancing Initiatives

In 2012, US President Barack Obama initiated a foreign policy shift to look-east, or "Pivot to the Pacific" or "rebalancing" toward the Asia-Pacific to deepen U.S. credibility in the region.[5] To neutralize rising China, the USA began establishing military, economic, and diplomatic ties with adjacent countries. It already has significant naval and air presence in Guam and Japan. It has defence cooperation with the Philippines and Singapore. The presence of American military in Afghanistan, Uzbekistan, and Tajikistan; strengthened ties with South Korea and Japan; effort to woo India and Vietnam is critical to the US 'Pivot to Asia Strategy'. The USA, Japan and Australia are often labelled as 'little NATO against China'. The USA is prioritizing the Pacific theatre with increased military presence and ultimately 60 per cent of the US Navy will be based in the Pacific. Meanwhile China is aggressively developing and deploying newer ships, AWACS and fifth generation fighters to protect islands in the SCS. The USA needs India to contain China. India also wants to manage an increasingly assertive China. To counter China's attempts to encircle it,

India has taken initiatives in the Indian Ocean neighbourhood to befriend Sri-Lanka, Bangladesh, Myanmar and also set up friendly bases at Seychelles, Mauritius, Tajikistan, and strengthen military presence in the Andaman and Nicobar Islands. India has increased its military and economic ties with ASEAN countries, Japan and Australia. Delhi is supporting Vietnam on training on Kilo Class submarines and Su-30s, and has offered anti-submarine torpedoes and BrahMos cruise missiles.

PLAAF's Regional Strategy

As per China's Military Strategy, May 2015,[6] in line with the strategic requirement of building air-space capabilities and conducting offensive and defensive operations, the PLAAF will endeavour to shift its focus from territorial air defence to both defence and offense and build an airspace defence force structure that can meet the requirements of informatized operations. The PLAAF will boost its capabilities for strategic early warning, air strike, air and missile defence, information counter-measures, airborne operations, strategic projection, and comprehensive support.

At the strategic and tactical levels, China's air power can now achieve a variety of effects. Though their current concentration is on operations on the eastern sea board in the SCS, the same platforms will be used against India. China wants to exploit the advantage of using its tactical/strategic missile force, which is easier to use for an offensive than defend against. The PLAAF plans to move the forward edge of the battle into enemy territory. It will use air offensive to keep the IAF head-down. Like the IAF, the PLAAF has switched to net-centric offensive air defence and greater reliance on integrated attack. China's ambition is to build its airpower like the USA for an asymmetric advantage. The scale of production saw the PLA's fourth and fourth-plus generation fleet size grow to at least 900-1,000 aircraft, crossing a symbolic milestone where more than half of their fighter fleet can be considered to be fourth generation or above.[7] Of greater concern also is the offensive capability in terms of PGMs and surface-to-surface missiles.

In the last two decades, China has made great strides in developing airpower capability, albeit, it remains work-still-in-progress. The introduction of fourth-generation fighters, bombers, unmanned aircraft and advanced cruise missiles has transformed it from a defensive force to one that can project Chinese power throughout Asia and its eastern Pacific sea board. The PLAAF's new capabilities will enhance Beijing's ability to project power. Of concern is the new long-range bomber that would be much more modern and exceed the range and capabilities of the H-6K which recently began flying with LACMs. In 2016, China and Ukraine agreed to restart production of the world's largest transport aircraft, the AN-225, which is capable of carrying a world-record payload of nearly 254 tons. On 25 March 2020, the aircraft commenced a series of test flights in Ukraine. China expects the first AN-225 to be delivered and operational by end 2020. If used by the military, this capability would facilitate the PLA's global reach. Growing PLA mission areas and enhanced presence abroad may lead to an increase in demand for the PLA to protect China's overseas interests and provide support to Chinese personnel.[8]

Road-Rail Infrastructure in TAR

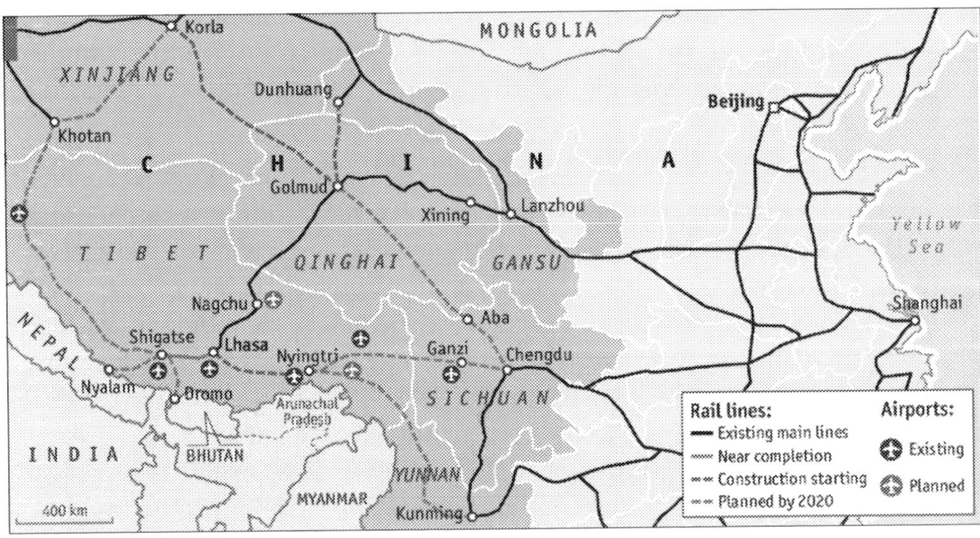

Infrastructure in TAR. *Map Source: The Economist 2014*
(Roads planned for 2020 have since been completed)

India and China share a 3,488-km long disputed border, 1,597 km in Eastern Ladakh, 545 km in the middle sector of Himachal Pradesh and Uttarakhand, 220 km in Sikkim and 1,126 km in Arunachal Pradesh.[9] China's 3,105 km Western Highway from Lhasa-Kashgar connects Xinjiang to Tibet and is a two-way, black top highway with a capacity of 3,200 tons per day. The 2,122-km Central highway connects Xining (Qinghai province) – Gormo – Lhasa. This is the main highway and the life line of Tibet with over 80 per cent of goods and passengers being transported. The 1715-km

Eastern Highway connects Chengdu to Ngiti (TAR) and has a capacity of 3,200 tons per day, and is effectively connected to Lhasa. The 1,142-km Qinghai to Tibet railway line is an engineering marvel as it goes through permafrost terrain at heights of 4,000 to 5,000 metres. It has the capacity of moving eight passenger trains per day and 5 million tons of goods. Link lines connect the railway network to areas close to the Indian border along the LAC. China is also considering a high-speed Trans Karakoram railway link from Kashgar to Havelian in Pakistan and onwards to Gwadar, which will go over the Khunjerab Pass and PoK area of Gilgit region. China has connected all the 39 major mountain passes touching India. The yawning gap in the quantity and quality of India and China's infrastructure near the LAC has multiple implications. China has nearly 12 divisions in the TAR and can rapidly mobilise up to 30.

Chinese Airfields in TAR

China has constructed 14 major airbases, and a number of ALGs and numerous helipads on the Tibet Plateau.[10] At least five airbases are already supporting Su-27 and Su-30 MKK fighter operations. China has upgraded the main air bases in TAR including Gangger, Pangta, Linzhi, Naqu, Saga.[11] There are two airfields in Lhasa. An additional four are being readied which can be made operational quickly. Many Chinese airfields in the TAR that face India are being upgraded with longer runways, hardened blast pens for housing operational aircraft and other operational assets. Hotan, Hoping and Shiquanhe, after upgrading, will facilitate better payload (weapon) carrying capability including in adverse weather operations. Many of these airfields are at an average altitude of 4,000 metres above sea level. Altitude affects performance of aircraft. Planes operating out of this altitude have to trade-off with fuel and payload. Besides there are another 20 airfields in the region, which can be further upgraded for military activity. While most airfields in TAR are used by civilian aircraft, Gonggar at Lhasa is the primary operating base for undertaking combat operations in Tibet. The PLAAF has begun developing infrastructure and operating from dual-use airfields at Ngari Gunsa, Nyingchi Mainling, Qamdo Bamda, Shigatse and Nagqu Dagring in the WTC. Of these, Shigatse is being converted for all-weather operations like at Gonggar, followed in the future probably by Ngari Gunsa. More than normal flying activity was observed over Lhasa-Gongga and Hotan airfields during the IAF's Gagan Shakti exercise. They flew missions to collect intelligence. The PLAAF can also use Kunming airfield, east of Myanmar, but at a much lower altitude. Additional ALGs have been constructed for all-weather operations of UAVs. Reportedly, new airfields are being constructed at Tashong, Shache, Loulan, Tumshuq, Qiemo and Fuyun.

Controlling Water: Strategic Dams in TAR

The Himalayas are South Asia's principal source of water. Nearly 47 per cent of the world's population in 10 countries depends on Tibetan water for its sustenance.[12] China can use water as a strategic commodity and as a tool for energy and economic diplomacy with neighbours. China has many upstream river projects. A barrage near Tsamda gorge near Guge kingdom could disturb the flow of the Sutlej and enable China to control and regulate the flow of water into India.[13] Similar actions may be taken in the Lohit (Zayul Chu), Subansiri and Indus, amongst others. India needs to have a strategy against such exploitation.

India's China Border Infrastructure Strategy

Although many years behind China, India has begun developing infrastructure in regions bordering China. India shares 3,488 km of border with China that runs along the States of Jammu & Kashmir (1,597 km), Himachal Pradesh (200 km), Uttarakhand (345 km), Sikkim (220 km), and Arunachal Pradesh (1,126 km).[14] The area is characterized by high altitude terrain and thick habitation which have resulted in inadequate development of infrastructure in these regions. The Indo-Tibetan Border Police Force (ITBP) guards the India-China Border. It has established 176 Border Out Posts (BOP) along the Line of Actual Control. The Government is undertaking construction of 73 roads of operational significance along the Indo-China border. The BRO is constructing 63 of these 73 roads. They have constructed 75 per cent of the entire length. The BRO also plans to complete about nine more such roads by March 2021. It has set a target to complete 61 of the 73 India-China Border Roads (ICBR) in three years.[15] The BRO is also constructing 125 strategic bridges in different sectors of the LAC, including Ladakh and Arunachal Pradesh. There are plans to construct 200 more bridges along this border. There are also plans to build the Arunachal Frontier Highway. This 2,000-kilometre-long road will follow the border between India and China. There will be another Trans-Arunachal Highway through the middle, and the Arunachal East-West corridor in the foothills along the Assam border. This infrastructure will help India to upgrade its military posture vis-a-vis China from 'dissuasion to deterrence'. Connectivity will help reinforce defensive formations and possible offensives.

New Indian Military Infrastructure and Formations Opposite China

The Indian Army has raised a new mountain strike corps (XVII) at Panagarh.[16] It has been built as a quick reaction force and as a counter offensive force against China to launch a quick counter-offensive and grab territory inside the TAR to strengthen India's position in the ensuing negotiations. The IAF will position a squadron of C-130 aircraft at Panagarh.[17] It will have to support the Indian Army's defensive formations to ensure that the Chinese cannot advance into Indian territory. India's eight ALGs in Arunachal,[18] at Walong, Mechuka, Vijoynagar, Tuting, Passighat, Ziro, Tawang and Aalo have already been upgraded and modernised. The IAF's Sukhoi-30MKIs and the advanced trainer Hawk Mk 132 carried out operations from six civil airports in Dimapur, Imphal, Guwahati, Pasighat, Andal and Kolkata in October 2019.[19] IAF fighters had earlier carried out trial landings at Lengui airport in Manipur. Similarly, Lilabari Airport in North Lakhimpur in east Assam has 7,500 feet runway and can be operationalised for fighters if required. Currently, India is trying to upgrade Lilabari for UAV operations. Many fighters, including LCAs, have operated from Leh airfield in Ladakh. The IAF's mighty C-17 Globemaster aircraft has landed at Mechuka ALG, just 29 km from the Sino-Indian border.[20] The IAF's C 130J-30 super Hercules made a historic landing at Daulat Beg Oldie (DBO), on August 20, 2013. The DBO is the highest airstrip in the world at 16,614 feet (5,065 metres) in the Aksai Chin area.[21]

PLAAF Air Strategy in Western Theatre

The Chinese Western Theatre Command (WTC) covers the entire Indian border. The WTC Theatre air forces[22] include the 37th Fighter Division at Urumqi, Xinjiang, which has one fighter brigade with J-8Hs; one fighter brigade with J-11A/Bs; one attack brigade with JH-7A and 2 SAM brigades. Lanzhou Base has one fighter brigade with J-11ABs; one fighter brigade with J-7s; one fighter brigade with J-7Es; one fighter/attack brigade with J-16s; and one SAM brigade. The 4th Transport Division at Qionglai, Sichuan, has one transport regiment with Y-8/Y-9; one transport regiment with Y-7; one transport regiment with Mi-17V-5/Y-7/Y-20. The

Lhasa base has one SAM brigade. There is also one Search and Rescue (SAR) regiment.

Chinese build-up in the TAR is significant. The growing capability of PLAAF in the TAR helps the economic development of the region, and capability of rapid deployment of forces for power projection across the border. In 2016, China elevated the rank of Tibet's Military Command which looks after security along its border with India, suggesting a higher level of military combat mission readiness.

PLAAF operations in TAR are now nearly throughout the year with the presence of J-10, J-11 and SU-27 aircraft on deployment or exercises. PLAAF has carried out weapon firing trials at high altitude and integrated exercises. Thirty-two J-10 aircraft were seen positioned in TAR at Lhasa and Hoping airfields at a given time. TAR thus has a significant might of PLAAF front-line air presence. PLA conducted live fire exercise with 'new artillery guns' at heights between 3,700 and 4,800 metres in TAR. The nuclear missile base in Quinghai province could be meant to target India. In addition to the five fully operational airfields (Gongar, Pangta, Linchi, Hoping and Gar Gunsa), Kashgar, located 600 km north of Srinagar, is an important operational airfield. The PLA can amass upward of two divisions at their launch pads along the border in just 20 days compared to the over 90 days it took earlier.[23] The limitations of high altitude operations with combat aircraft of yesteryears stands diluted to a large extent with the induction of modern aircraft like SU-27s and J-10s with aerial refuelling capability. Strike aircraft can thus take off with a combination of minimum fuel and maximum weapon load to be refuelled in the air. SU-27s and J-11 thus have the capability to strike deep inside Indian territory.

Former Indian Air Force Chief Air Chief Marshal B.S. Dhanoa said during a briefing on exercise 'Gagan Shakti' in April 2018, that a significant increase in Chinese aircraft and aircrew had taken place in the TAR as compared to other military regions. He added that China's deployment of significant Sukhoi-27 and J-10 fighter aircraft for continuous operations during winter months gives them a "year round capability."[24] Although no frontline PLAAF fighter elements are based there, regular rotation deployments to Tibet are an important part of the training doctrine. China is likely to develop additional all-weather strips in the future. In recent months, a quick reaction element of J-11A/B or Su-27SKs is always available at the Gonggar airbase at Lhasa.

The Air Defence of Chinese Western Theatre

The PLAAF Air Defence component in the WTC comprises of two Air Regiments of J-11A/B/BS (98th AR and 111th Brigade) and one mixed Regiment of Su-27SK and J-11Bs (16th AR).[25] By 2020, it was expected that there will be a minimum of three Operational Readiness Platform (ORP) detachments (with at least eight aircraft) available all through the year in Tibet, with probably a full-time squadron deployment at Shigatse by 2025. Other aircraft in the Chengdu region include J-10 (one regiment), SU-27 UBK (one regiment), J7B (two regiments), besides aerial tankers (IL 78), MI 17 V7 helicopters and J-6, J-7, and SU-30 aircraft. In adjoining Lanzhou, there are two fighter and one bomber divisions. FRA, Airborne Early Warning (AEW) aircraft and a strong ground-based air defence and Ballistic Missile Defence (BMD) cover centred around S-300/400 and a variety of Short, Medium and Long Range SAMs adding to the might.

PLAAF Strategy in TAR

The PLAAF has stocked a large number of Battlefield Support Missiles (BSM) in Xinjiang and Aksai Chin for possible use in CAO missions against Indian airfields and ALG across the Line of Actual Control.[26] For interdiction operations on the supply lines of India's forward deployed ground forces in both Eastern Ladakh, Himachal Pradesh and Uttarakhand, the PLAAF is likely to employ a combination of massed fire assaults from heavy calibre MBRLS along with Battlefield Air interdiction (BAI) sorties employing SU-30 Mk-2 and J-10 aircraft, duly supported by AEW platforms. In order to rehearse for the above contingencies, adequate training is being carried out regularly. Joint expeditionary army-air force live firing exercises are being conducted with increasing frequency and scale. A variety of weapons like rockets, BVR Missiles, Laser Guided Bombs, etc., are fired. Against India, China's focus would be in Arunachal Pradesh area. They will try to take advantage of the limitations of radar cover in the mountains. It shall be their endeavour to put down the main Indian fighter airbases.

China's Likely Tactical Moves against India

The Tibet Military Command will be responsible for operations against India, at least in the Arunachal Pradesh area, training forces for specialized high-altitude mountain warfare and long-range mobility for such a contingency.[27] The Xinjiang Military Command may also be elevated in the future. Both commands are under the newly created Western Theatre Command. The lack of cross-terrain mobility limits the ability of ground forces to conduct penetrating or outflanking operations against enemy forces. PLA publications stress airmobile landings in the enemy rear area to overcome the restricted terrain and enemy defensive positions. Special operations forces available to the WTC would represent highly qualified units to operate in the enemy rear area to disrupt operations and attack vulnerable lines of communications. The high-altitude reduces aviation performance and lift capabilities, and increases maintenance requirements on equipment in general, although the thin air increases the range of projectiles and shrapnel. Weather conditions would mostly limit air operations to June through November. Cold high-plateau regions place increased requirements on engineering and support operations, and the thin air is difficult for the troops even after acclimatisation. As an independent campaign, a joint fire strike could represent punitive strikes against key Indian targets. A joint fire strike campaign is a long-range precision strike by long-range rocket, missile and air forces with the objective to destroy important enemy targets, paralyze the enemy's operational systems, weaken the will to resist and destroy war potential, or create conditions for other operations.

The primary border areas under dispute are the Chinese-occupied Aksai Chin in the west, and Arunachal Pradesh in the east controlled by India. As has been seen during the Doklam incident, China was preparing for the National Congress of Communist Party in November 2017, and it was a 'Show of Power' time for various competing power centres. Many of China's border stand-offs have been mostly posturing. With the Trump Administration becoming conservative in trade policies and European economies shrinking, India remains an important market for China with over US$ 100 billion annual trade. They should not like to 'hurt' the goose that gives the golden eggs. India is now a closer ally of the USA and the Americans and Russians may prevent China from entering into a full war. India has been engaging China in one-to-one meetings between heads of the two states. Confidence Building Measures (CBM) are also in place. Continuing to expand military to military engagements, the Eighth Edition of the India-China joint training exercise 'Hand-in-Hand-2019' was held at Joint Training Node, Umroi, Meghalaya, on 07 December 2019. The PLA contingent from the Tibet Military Command and the Indian Army contingent comprising one Infantry company along with supporting staff were participating in the 14-day joint training exercise.[28] The exercise aimed to enrich both the contingents from each other's experience in counter-terrorism operations. Besides counter-terrorism operations, discussions on Humanitarian Assistance & Disaster Relief (HADR) operations were also conducted as part of the exercise. The exercise was previously held in December 2018 at Chengdu, China. Company size contingents of the Indian Army's 11 SIKH LI and a regiment from the Tibetan Military District of the People's Liberation Army participated in the exercise.[29] There have been military-to-military exchange visits at higher levels. A military delegation led by then Northern Commander Lt Gen Ranbir Singh made a five-day visit to military establishments in Beijing, Chengdu, Urumqi and Shanghai in January 2020. They interacted with the General Commanding WTC at his headquarters at Chengdu.[30]

Doklam is an area of particular concern to Indian planners, as a small southward thrust of Chinese territory looks like a dagger pointed toward one of India's most vulnerable geographic points—the so-called 'Chicken's Neck' or the Siliguri Corridor that runs between Nepal, Bhutan, and Bangladesh and connects the main portion of Indian territory with its north eastern provinces that lie to the east of Bangladesh.[31] Chinese Generals would like to cut off India's north-east region. The IAF does have some important air bases around the Siliguri Corridor. If China were seizing a key piece of terrain, it might seek to avoid engaging Indian forces. If, however, China were seeking to teach India a lesson as it did in 1962, it might instead target a vulnerable Indian unit to cause attrition or take prisoners. If China were seeking a broader reordering of the political-military status quo in the region, its approach might be more deliberate and expansive.

Nevertheless, China's approaches are likely to have some common themes.[32] Prior to any potential conflict, China would likely spend weeks, months, or even years gathering and developing information on its targets and the disposition of the Indian forces it might face. Indian forces generally outnumber Chinese forces along the border. China would likely seek to offset this disadvantage by using covert movement or deception—such as acting under the guise of an exercise—to concentrate forces against a weak point in India's defence perimeter. China would move quickly to seize key terrain such as passes or high ground with good fields of fire. Then, China would rapidly reinforce its position from other theatres using its road and rail networks in the region. At the same time, China would probably use firepower (likely artillery, given the altitude and terrain) and combat engineering to prevent a counterattack while at the same time launching feints against other points along the border to confuse India's response and prevent an effective counterattack. Having established a fait accompli, China would present India with an unpleasant dilemma of counterattack and potentially escalate a conflict against a superior foe, or accept a loss of territory and the political ramifications that follow.

The mountainous terrain significantly limits the effective ranges of sensors and communications networks.[33] At the same time, the light units India might put forward on the border make poor targets for long-range precision weaponry. China could use long-range precision fire to strike airfields, roads, railways, and other key infrastructure that India's armed forces rely on to respond to an incursion. However, given the number of potential Indian targets, a major strike campaign will likely not be worth the cost or the risk of escalation from China's perspective. It is far more likely that China would use its precision fire to disrupt and degrade forces near its offensive, while also isolating the battle space.

India seeks to defend all its terrain with light forces and to reinforce this forward posture quickly, while also counterattacking into Chinese territory with heavier manoeuvre forces to seize terrain to compel conflict termination or for post-conflict negotiation. But defending the entire border may be a difficult proposition.[34] India should exploit its defensive posture and geographical advantages to draw Chinese forces into disadvantageous encounters and stretch China's lines of communication. India should use unconventional forms of manoeuvring to attack the PLA's critical systems, disrupt their operations, and impose costs on the Chinese regime. India should resort to offensive cyber-attacks to disrupt networks; camouflage, concealment, and deception; jamming/dazzling the sensors. Instead of combined-arms manoeuvres to seize territory on the Tibetan plateau for post-conflict trades—which might open up Indian forces to withering air and missile strikes—India should hold its manoeuvring forces in the mountains and use unconventional warfare and offensives. It is better to cede some territory, but it would impose considerable costs on the PLA and the Chinese regime, while putting Indian forces in a strong position to counterattack to regain losses.

The Chinese could also create for themselves a favourable balance of forces along certain points on the LAC and then even sustain gains by extending their own border infrastructure very short distances into Indian territory, relatively quickly.[35] In at least two places in Ladakh, Chinese border roads extend till points that lie within what India perceives to be territory controlled by it. Typically, the Chinese 'road head' is either right up to the LAC or is just a few hundred metres short of it while the Indian 'road head' could be anywhere between several hundred metres to several kilometres short of the LAC, although this is changing now. In any case, the Chinese have built motorable tactical roads from their Western and Eastern highways in TAR up to all the 31 passes (eleven in the Western Sector, five in the Middle Sector and 15 in the Eastern Sector) that are of military significance along the LAC. In addition, various border 'laterals' of low classification also exist just south of subsidiary axes to the main tactical roads that may be used for an advance.

India's focus in recent times has been on building infrastructure that makes it easier for the Indian military to access even rather remote mountainous frontiers, such the easternmost parts of Arunachal Pradesh, where in places such as Kibithu in Anjaw district the Chinese have been 'asserting' themselves.[36] Indeed, Kibithu, is believed to be one of the likely targets for a Chinese attack in the event of hostilities. Other targets of value for the Chinese would of course be Tawang in western Arunachal Pradesh where the sixth Dalai Lama was born, Chaglagam in Arunachal Pradesh's Lohit district, the Depsang Plains and Chumar in Ladakh (where PLAGF troops intruded during Xi Jinping's 2014 visit) and possibly Barahoti in Uttarakhand's Chamoli district.

Probably the target of greatest value for the Chinese is Tawang, since they specifically claim the concession of the Tawang tract to them as a pre-requisite for settling the India-China border dispute. The Chinese also occupy key features opposite Indian forces deployed in Tawang which allows them to peer into Indian territory. Nevertheless, the Indian Army has positioned enough forces for the defence of Tawang.[37] They have fortified Tawang with numerous bunkers, dugouts for artillery, underground shelters for pre-positioned war material, camps for acclimatizing troops before deployment to higher altitude posts on the LAC and small live fire training areas to ensure continuous readiness. Of course, there is still essentially a single axis road that connects Tawang to the plains below and is now being re-carpeted with long lasting concrete by India's Border Roads Organization, which also intends to construct a tunnel to bypass the Sela Pass. An ALG is also being constructed in Tawang.

Aerial Threat to India

China's vast network of highways and rail connectivity in TAR gives it the advantage of quick build up and sustenance. Any attack on India would include attacks by conventional ballistic missiles of the PLARF. They will target airbases, AD command and control centres, SAM sites and Electro Magnetic (EM) emitters. The SSMs will be followed by manned bombers (H-6) and front-line fighter aircraft using cruise missiles to target EW sites, SAMs, radars and EM emitters that may be still operational. Fighters could also launch laser-guided and satellite-guided bombs. The limitations of high altitude operations with older combat aircraft stands diluted largely with the induction of modern aircraft and other 'enablers' like FRA and AEW aircraft. Simultaneously, PLAAF is likely to embark upon an offensive air campaign to seize air superiority or a favourable air situation. The next set of targets could include Indian SSM launch pads. Other targets would be the Indian naval bases and naval forces at sea. Lastly, they may attack the seat of power, economic targets, and core-sector targets like oil dumps and refineries.

China's growing build up in TAR is indicative of their campaign strategy. The capability build up in the Tibet region achieves the dual aim of economic development and offensive posturing against India. Improved airfield infrastructure and road and rail networks support rapid deployment of forces. The entire Sino-India border is covered by China's Western Theatre Command, which has a geographical area of responsibility as big as India. The relatively low altitude airfields in the Southern Theatre Command are just across Myanmar. In case of an eventuality, PLAAF will over fly Myanmar to strike India's eastern airfields, with or without their permission.

The PLAAF has stocked a large number of aerial weapons and SSMs in the region for air operations against India. It will fly BAI missions mostly with J-10 aircraft and J-11, duly supported by AEW platforms. Adequate training is being carried out regularly. J-11s and SU-27s have been conducting air superiority missions in all weather conditions. The J-11s have also been doing night operations. The regiments of the PLARF deployed in the Tibetan Plateau regularly conduct 'live fire' exercises.

The Air Operations

Most Chinese airfields in Tibet have long runways to cater for all types of operations at high altitude. Ngari Gunza airfield,[38] not too far from

Leh, has a 4,500 m (14,764 feet) runway. Air Operations have been mainly by J-10s and J-11s. Total yearly duration for activation has gone up despite a significant period of being snowbound. The PLAAF will certainly move in SU-30 class of aircraft. China is likely to take the initiative of first strike to neutralise critical airfields in India's Ladakh and the North East for a possible localised war. Chinese strategy will be to attack critical IAF airbases with a mix of surface fired missiles and PGM carrying fighter-bombers. The IAF will have to prepare to defend its airfields. It has been positioning state-of-the-art fighter aircraft at many of its 25 airbases from Leh and Nyoma in Ladakh to Tezpur and Chabua in the east along the China border.

Winning the air war is critical for Indian Army operations. Interdiction of critical bridges on the Chinese road and rail network will thwart PLA movements and result in maximum dividends. IAF air bases, being in the plains, give take off payload advantage vis-à-vis PLAAF operations from high altitude TAR airfields. The IAF will have to make repeat attacks on the few Chinese airfields in TAR to keep them neutralised. It is building logistic infrastructure and hardened aircraft shelters at all its airbases. Top end IAF fighters with combinations of PGMs and BVR missiles supported by AWACS will launch offensive and defensive missions. Indigenous BrahMos Block III cruise missiles with steep dive capability could help knock off targets deeper across the LAC. Inter-theatre and intra-theatre movement of troops can now be greatly supported by C-17s, IL-76s and C-130s. Heavy lift helicopters will help move critical army enforcements inter-valley in the mountains. The AH-64E Apaches could support anti-tank operations in North Sikkim and Ladakh. Myanmar may allow overflights to the PLAAF during a conflict with India; Nepal may remain neutral; Bangladesh is also likely to be neutral. India would have to factor in the possibility of Pakistan allowing the PLAAF to use Gilgit, Skardu, Chilas, Chitral and Muzaffarabad airbases.

Inferences and Options for India

It is unlikely that China will embark even on a limited war with India with the present state of the PLAAF but a decade or more down the line, the PLAAF would have built up a greater capability lead over India, both in quality and quantity. Currently, the two sides can be considered evenly matched across the Himalayas, and, if at all, India has the edge in air power. By then, China aspires to attain a conventional capability to project itself under a nuclear deterrence umbrella. It may decide to seek a solution to its claims in Arunachal Pradesh, through a limited war. China has much better ground infrastructure in the Sino-Indian border region for possible operations against India. China is likely to run a campaign of punitive joint-fire strikes against key Indian targets, using long-range precision strikes by long-range rocket, missile and air forces to destroy important targets, paralyze operational systems, weaken the will to resist and destroy war potential, or create conditions for other operations.

Defence analyst Rick Joe has assessed the air domain of the Taiwan invasion.[39] Employment of air power in Tibet may, in some cases, be on similar lines. PLAAF air strikes will be conducted in conjunction with PLA Rocket Forces and coordinated air and missile strikes against high value targets. PLAAF's medium-large AEW&C aircraft such as KJ-500, KJ-200 and KJ-2000 aircraft would maintain persistent 24-hour orbits to monitor enemy air activity and relay information to PLAAF aircraft, air defences, and build overall situational awareness. Two or more ELINT and SIGINT aircraft would also operate at standoff range within Chinese airspace alongside two or more Y-8G/Y-9G standoff ECM aircraft operating against enemy air defences, early warning radars and fighter aircraft. The pace at which the PLA will seek to apply air power—especially for air superiority and strike missions—will greatly depend on the success of preceding long range conventional strikes from SRBMs and LACMs, as well as the effectiveness of PLAAF AEW&C aircraft, EW/ECM aircraft, and ISR aircraft. The PLAAF would conduct follow-on operations. Its quantitative fighter advantage will almost certainly be further compounded by the much larger advantage in AEW&C aircraft, standoff EW/ECM aircraft, and ELINT aircraft. The PLAAF not only enjoys a significant advantage in airframe numbers but also overall system capability, size, and endurance.

India will need to build effective surveillance capability in the TAR region using airborne and space based sensors. It needs to have a seamless capability for all-weather air defence, Counter Air Operations (CAO), interdiction and Counter Surface Force Operations (CSFO) capability. India must quickly build the IAF fighter aircraft strength to the authorized 42 squadrons with all 4th generation plus fighters. India needs to get its act right for 5th generation aircraft or procure at least 3-4 squadrons. India will also have to tackle Chinese stealth aircraft. It must develop/acquire a suitable stealth aircraft, ARMs and cruise missiles. India's air defence

Surface-to-Air Missiles (SAM) must take on the Surface-to-Surface Missile (SSM) threat. There is also a need for system survivability from physical or electronic attack and redundancy. For India to keep the PLAAF at bay, there will be the need for capability to carry out effective surveillance of the TAR region using air and space sensors to create a comprehensive air situation picture. Surveillance gaps would have to be identified and plugged. The IAF's Integrated Air Command and Control System (IACCS) must ensure continuous coverage and response to any air threat in the region. AD must cover the full spectrum from LRSAM and MRSAM to SRSAM to defend against aircraft and missile attacks.

India's ICBM and ABM capability has grown with at least 110 nuclear warheads with missile ranges of nearly 8,000 km. In addition, the IAF's conventional strength and preparation levels are a deterrent for any Chinese misadventure. The PLAAF has unified air and space defence. India needs to also accelerate integrating space and air assets.

The PLAAF will have integrated joint operations with the PLARF in TAR, with the latter having at least two major missile bases with three missile brigades each in the Western TC.[40] PLARF could use conventionally armed missiles in the opening stages of any conflict to attack IAF airbases in addition to other targets such as C2 nodes, AD sites, etc., thereby making it a key enabler of air operations for the PLAAF. Alternatively, the PLARF could be used later in 'retaliatory mode', if the Chinese start facing reverses in various sectors. Indian military war-games always factor in a major conventionally armed missile attacks by the PLARF in various scenarios.

Another flashpoint is the Depsang Plains that abuts the Siachen Glacier and the all-important Karakoram Pass.[41] In 2013, the area witnessed a major incursion by the Chinese Army that led to a standoff which was defused only after the Indian Army managed to deploy sizeable forces there with the support of IAF Mi-17 helicopters since road connectivity was non-existent at the time. A few months later, the IAF demonstrated a C-130J landing at DBO in Depsang. Nevertheless, the area continues to be perceived as vulnerable by the Chinese because of its vicinity to their Western Highway. The PLA's Western TC Air Force does have three airfields in the vicinity at Kashgar, Shiquan and Hotan, wherefrom PLAAF fighters can take-off with sizeable payloads. India has deployed an entire armoured brigade in Eastern Ladakh, with two T-72 regiments already operational. DBO is also now connected via a new black top road. Besides providing a stouter defence of DBO, the armoured brigade in Eastern Ladakh could also be used to spearhead an attack towards the Western Highway passing through Aksai Chin via the Chushul-Demchok axis. There are ALGs at Fukche and Nyoma in Eastern Ladakh with Nyoma also hosting a T-72 regiment.

IAF aircraft will clearly have an endurance advantage in any air war over Tibet.[42] Moreover, the adverse meteorological conditions over Tibet are difficult to negotiate at the best of times and it remains to be seen how well the PLAAF would be able to cope with it during a major air campaign, especially given the limited infrastructure at most TAR airfields. In sharp contrast, all major IAF airbases with relevance to the LAC have been upgraded as part of the modernization of airfield infrastructure (MAFI) project which has seen the installation of air traffic control and landing aids that makes them capable of hosting aircraft for day/night all-weather operations.

When fully raised, the Mountain Strike Corps (MSC) will have four high-altitude infantry divisions, two independent infantry brigades, two armoured brigades, 30 new infantry battalions and two Para Special Forces (SF) battalions.[43] Apart from M-777s, it will also feature a strong aviation component that will eventually include AH-64E Apache attack helicopters and Hindustan Aeronautics Limited (HAL) Light Combat Helicopter (LCH). India is also inducting a regiment of the Brahmos Block III land-attack cruise missiles (LACM) into its NE which has 'trajectory shaping' capability, that allows it to attack targets buried near mountain sides. In the future, a new SRBM called the Pralay will also be brought into this sector. The idea is to level the playing field for the IAF via a symmetric counter to the PLARF's conventionally armed missile deployments.

At current strength, the IAF will find it tough to take on both the PLAAF and PAF. It requires at least 50 combat squadrons for a possible two-front war. India's defence outlay at $ 61 billion is too little for defence modernisation, leave alone the IAF bridging the combat squadron gap. The defence outlay has to go up from the current 1.5 per cent of the GDP to 2.5 per cent. For a two-front war, the IAF will require to double the number of AEW&Cs, heavy lift aircraft of C-17 and IL-76 class, and FRA. India needs to succeed in 'Make in India' in defence. The IAF has miles to catch up. The world is watching with concern as the once sleeping Dragon begins to spit fire. India is the most affected regional player. The time to act is now.

REFERENCES

1. Blake Stilwell, 6 brutal facts about 'flying the hump' — the deadly World War II air route over the unforgiving Himalayas, *Business Insider*, June 15, 2018, https://www.businessinsider.com/6-brutal-facts-about-flying-the-hump-wwii-airlift-route-to-china-2018-6?IR=T
2. Sakamoto Shigeki, The Senkaku Islands as Viewed through Chinese Law, The Sasakawa Peace Foundation, September 12, 2016, https://www.spf.org/islandstudies/research/a00017r.html
3. Maninder Dabas, Here Is All You Should Know About 'String Of Pearls', China's Policy To Encircle India, indiatimes.com. June 23, 2017, https://www.indiatimes.com/news/india/here-is-all-you-should-know-about-string-of-pearls-china-s-policy-to-encircle-india-324315.html
4. Jonathan Woetzel, Shanghai; Jeongmin Seong, Shanghai; Nick Leung, Hong Kong; Joe Ngai, Hong Kong; James Manyika, San Francisco; Anu Madgavkar, Mumbai; Susan Lund, Washington, DC and. Andrey Mironenko, Sydney. China and the world: Inside the dynamics of a changing relationship, McKinsey Global Institute, p. 1, July 2019, https://www.mckinsey.com/~/media/mckinsey/featured%20insights/china/china%20and%20the%20world%20inside%20the%20dynamics%20of%20a%20changing%20relationship/mgi-china-and-the-world-full-report-june-2019-vf.ashx
5. Mark E. Manyin, Stephen Daggett, Ben Dolven, Susan V. Lawrence, Michael F. Martin, Ronald O'Rourke and Bruce Vaughn, Pivot to the Pacific? The Obama Administration's "Rebalancing" Toward Asia, Congressional Research Service Report, March 28, 2012, https://fas.org/sgp/crs/natsec/R42448.pdf
6. China's Military Strategy, USNI News, May 26, 2015, https://news.usni.org/2015/05/26/document-chinas-military-strategy
7. Rick Joe, China's Military Advancements in the 2010s: Air and Ground, *The Diplomat*, February 05, 2020, https://thediplomat.com/2020/02/chinas-military-advancements-in-the-2010s-air-and-ground/
8. CHINA MILITARY POWER Modernizing a Force to Fight and Win, US Defence Intelligence Agency Report 2019, p. 83, https://www.dia.mil/Portals/27/Documents/News/Military%20Power%20Publications/China_Military_Power_FINAL_5MB_20190103.pdf
9. Lt Gen Vinod Bhatia, CHINA's INFRASTRUCTURE IN TIBET AND POK - IMPLICATIONS AND OPTIONS FOR INDIA, CENJOWS, 2016, https://cenjows.gov.in/pdf/CHINAs_INFRASTRUCTURE_IN_TIBET_Inside_Layout.pdf
10. Ibid.
11. Sudhi Ranjan Sen, India Today, China upgrades air capability in Tibet, monitors Indian exercise, May 22, 2018, https://www.indiatoday.in/india/story/china-upgrades-air-capability-in-tibet-monitors-indian-exercise-1239082-2018-05-22
12. Lt Gen Vinod Bhatia, CHINA's INFRASTRUCTURE IN TIBET AND POK - IMPLICATIONS AND OPTIONS FOR INDIA, CENJOWS, 2016, https://cenjows.gov.in/pdf/CHINAs_INFRASTRUCTURE_IN_TIBET_Inside_Layout.pdf
13. Satish Kumar (Ed.), India's National Security, Annual Review 2014, p. 402, Foundation for National Security Research, New Delhi
14. MHA Website, https://mha.gov.in/sites/default/files/INDO_China_16022018.pdf
15. Shaurya Karanbir Gurung, In ramp-up, 75% roads on China border ready, *Economic*

Times, January 20, 2020, https://economictimes.indiatimes.com/news/defence/in-ramp-up-75-roads-on-china-border-ready/articleshow/73409778.cms?from=mdr

16. Pooja Mehta, Army's Mountain Strike Corps moves to its Panagarh headquarters for close watch on China, *Zee News*, March 01, 2019, https://zeenews.india.com/west-bengal/armys-mountain-strike-corps-moves-to-its-panagarh-headquarters-for-close-watch-on-china-2184505.html
17. Panagarh to be IAF's second hub for C-130J Super Hercules planes, *Economic Times*, (PTI) July 11, 2018, https://economictimes.indiatimes.com/news/defence/panagarh-to-be-iafs-second-hub-for-c-130j-super-hercules-planes/articleshow/48344173.cms?from=mdr
18. Deepshikha Hooda, IAF to have 7 operational Advanced Landing Grounds in Arunachal Pradesh in a month, *Economic Times*, July 13, 2018, https://economictimes.indiatimes.com/news/defence/iaf-to-have-7-operational-advanced-landing-grounds-in-arunachal-pradesh-in-a-month/articleshow/49809013.cms?from=mdr
19. Vishnu Som and Debanish Achom, IAF To Carry Out Fighter Ops From Civilian Airports In North-East, Bengal, *NDTV*, October 17, 2019, https://www.ndtv.com/india-news/indian-air-force-to-carry-out-fighter-operations-from-civilian-airports-in-north-east-west-bengal-2118248
20. IAF's C-17 Globemaster can now operate from remote Arunachal ALG, *Business Standard*, (PTI), November 4, 2016, https://www.business-standard.com/article/current-affairs/iaf-s-c-17-globemaster-can-now-operate-from-remote-arunachal-alg-116110400007_1.html
21. IAF C 130J-30 super Hercules Makes an Historic Landing at Daulat Beg Oldie (DBO), PIB Release, August 20, 2013.
22. *The Military Balance* Chapter Five: Russia and Eurasia,120:1,166-219, DOI: 10.1080/04597222.2020.1707966, p. 266, February 13, 2020, https://doi.org/10.1080/04597222.2020.1707966, https://www.tandfonline.com/doi/abs/10.1080/04597222.2020.1707966
23. Lt Gen (Dr) V.K. Saxena, Growing Muscle of PLAAF, Vivekananda International Foundation Occasional Paper – March 2016, p. 15, https://www.vifindia.org/sites/default/files/growing-muscle-of-plaaf.pdf
24. Shaurya Karambir Gurung, *Economic Times*, Significant increase in Chinese aircraft in Tibet Autonomous Region: IAF Chief B. S. Dhanoa, April 27, 2019, https://economictimes.indiatimes.com/news/defence/significant-increase-in-chinese-aircraft-in-tibet-autonomous-region-iaf-chief-b-s-dhanoa/articleshow/63929299.cms?from=mdr
25. Sameer Joshi, The Dragon's Claws, *Vayu Aerospace* and *Defence Review*, August 28, 2018, https://www.vayuaerospace.in/article.html?n=rafael—supporting-%E2%80%98make-in-india%E2%80%99&d=23
26. Lt Gen (Dr) V. K. Saxena (Retd), Growing Muscle of PLAAF, March 2016, https://www.vifindia.org/sites/default/files/growing-muscle-of-plaaf.pdf
27. Kevin McCauley, Snapshot: China's Western Theatre Command, China Brief Volume: 17 Issue: 1. January 13, 2017, https://jamestown.org/program/snapshot-chinas-western-theater-command/
28. Press Information Bureau, Government of India, Ministry of Defence, December 08, 2019, https://pib.gov.in/newsite/PrintRelease.aspx?relid=195593
29. Shaurya Karanbir Gurung, India, China to carry out major military exercise 'Hand in Hand', July 20, 2019, https://economictimes.indiatimes.com/news/defence/india-

china-to-carry-out-major-military-exercise-hand-in-hand/articleshow/70302866.cms?from=mdr

30. Successful China visit by military delegation paves way for future collaboration: Army. *The Times of India*, (PTI), January 11, 2020, https://timesofindia.indiatimes.com/india/successful-china-visit-by-military-delegation-paves-way-for-future-collaboration-army/articleshow/73205088.cms
31. Chris Dougherty, Force Development Options for India by 2030, Centre for North American Security (CNAS), October 23, 2019, https://www.cnas.org/publications/reports/force-development-options-for-india-by-2030
32. Ibid.
33. Ibid.
34. Ibid.
35. Saurav Jha, Duel In The Himalayas: How India And China Square Off, *Delhi Defence Review*, July 26, 2017. http://delhidefencereview.com/2017/07/26/duel-in-the-himalayas-how-india-and-china-square-off/
36. Ibid.
37. Ibid.
38. Ngari Gunsa Airport, Wikipedia, https://en.wikipedia.org/wiki/Ngari_Gunsa_Airport
39. Rick Joe, Anatomy of a Taiwan Invasion: The Air Domain, *The Diplomat*, April 03, 2019, https://thediplomat.com/2019/04/anatomy-of-a-taiwan-invasion-the-air-domain/
40. Saurav Jha, Duel In The Himalayas: How India And China Square Off, *Delhi defence Review*, July 26, 2017. http://delhidefencereview.com/2017/07/26/duel-in-the-himalayas-how-india-and-china-square-off/
41. Ibid.
42. Ibid.
43. Ibid.

11

India's Aerospace Options

"The Air Force will extend its reach from the sky to the space for defence of Chinese territory and to attack as well. We will improve the overall capability to strike at long distance targets with high precision, fight electronic or internet warfare with backup from space."

—**Xu Qiliang** (Former PLAAF Chief &
Current Vice Chairman, CMC)

"Every neighbouring state is an enemy and the enemy's enemy is a friend."

—**Kautilya**, *The Arthashastra*

As part of the 'Grey Zone' information war approach, China has been pumping stories into the public domain about its rapid armed forces modernisation and growing aerospace industrial capability. Its White Paper on defence focuses on equipment. The Indian armed forces continue to be manpower-heavy. As Chinese philosopher Lao Tzu said, "There is no greater danger than underestimating your opponent."[1] China's centralised leadership has set ambitious targets to acquire all major aerospace technologies by 2025[2] and might be able to equal U.S military spending in directly comparable cost terms by some point between 2030 and 2040.[3] In many indicators of the U.S.-China military scoreboard, China is fast catching up as per a Rand report.[4] Also, the aerospace sector is being pumped up with huge budgets. China is concentrating on game-changing technologies, AI, hypersonic, directed energy weapons, and long-range missiles. China's defence budget has been nearly four times that of India for the last two decades. China's defence research budget at nearly US$ 25 billion is many times more than India's. The cash rich country continues to build huge military infrastructure close to Indian borders and is extending its naval presence in the Indian Ocean. The Indian aerospace industry, despite successes in the space sector, is at least two decades behind. While India is still struggling with the 4th generation LCA production, China already has two stealth fighter aircraft flying, indigenous large transport aircraft, working on a stealth bomber, advanced helicopters, and many UAVs. In addition, the Pakistan-China strategic partnership-cum-alliance has to be contended with.

Under these circumstances, India needs a three-pronged approach. Firstly, accelerate aerospace systems development and production with clear central focus. Secondly, identify and concentrate on evolving force multiplier technologies and systems. Thirdly, create geo-political alliances for overall security in the region.

Counters to China's Precision Strike Capability

China's Second Artillery Force became the PLA Rocket Force (PLARF) on December 31, 2015. Additionally, the PLARF was elevated from an independent branch to the fourth military service alongside the PLA, PLAN, and PLAAF. It indicates the importance China puts on maintaining modern missile forces. According to the US-China Economic and Security Review Commission,[5] the PLARF "has at least 1,330 and potentially more than 1,895 ballistic and cruise missiles, which includes 1,000-1,200 short-range ballistic missiles, 75-100 medium range ballistic missiles, 5-20 intermediate range ballistic missiles, 50-75 intercontinental ballistic missiles, and 200-500 ground-launched land-attack cruise missiles." The PLARF's missile systems, coupled with the PLA's rapidly developing space and counter space platforms, have become critical components of China's emerging power projection capabilities. With the addition of new modernized missile classes and various satellite and counter-space capabilities, the PLARF is now capable of credibly deterring adversaries at intercontinental ranges and conventionally holding at risk adversary forces within 1,500 km of China. These changes are the result of major doctrinal modifications that started during the 1980s and fundamentally altered the PLARF's overarching mission as well as its position within the wider PLA.

The force fields advanced variants with improved ranges and accuracy in addition to more sophisticated payloads that can target land targets and naval ships. At the same time, the development of conventional short-range ballistic missiles (SRBMs), medium-range ballistic missiles (MRBMs), and land attack cruise missiles (LACMs), as well as its improving intercontinental ballistic missiles (ICBMs) means China must modify its strategy, and develop a wide range of additional command and control, and battle management capability. These include developing a new approach to deterrence, new approaches to war fighting, and a more sophisticated capability to target enemy forces over the horizon while coordinating such attacks with joint PLA forces.[6]

China has over-the-horizon (OTH) targeting capability with sky wave and surface wave OTH radars, which are used in conjunction with reconnaissance satellites and support long-range precision strikes.[7] The PLA can field 200-300 air or ground-launched Land-Attack Cruise Missiles (LACM) such as YJ-63, KD-88, CJ-10, and the CJ-20 for standoff precision strikes. China's CM802AKG LACM, an export system that can strike both land and ship targets from fighters or bombers, is also with Pakistan. The PLAAF has an inventory of tactical air-to-surface missiles (ASM) as well as precision-guided munitions including all-weather, satellite-guided bombs, anti-radiation missiles, and laser-guided bombs. China is developing smaller-sized ASMs for its UAVs. It deploys a wide range of advanced Anti-Ship Cruise Missiles (ASCM) which are deployed on the majority of China's ships as well as many aircraft. China has also developed the long range supersonic YJ-12 ASCM for the H-6 bomber. China has integrated an indigenous version of the Russian Kh-31P (AS-17) Anti-Radiation Weapon, known as the YJ-91, into its fighter-bomber force.

China's strategy is to first use its missile forces for Suppression of Enemy Air Defence (SEAD). Nearly 500 of its increasing precision missile inventory could be deployed or used against India. The IAF has around 16 units of relatively old Russian Pechora S-125 SAMs (max 35 km range) which India has been trying to upgrade indigenously into a digital system so as to make it compatible to the Integrated Air Command and Control System (IACCS), and extend their life.[8] The indigenous Akash (SAM) can engage targets such as aircraft, cruise and ballistic missiles up to 30 km range and 18 km altitude. Each battery can track up to 64 targets and attack up to 12 of them and defend an area of 2,000 sq. km. Newer variant Akash-NG will have a range of 50 km.[9] The IAF has 8 squadrons, with 125 missiles each, and 7 more units have been ordered. The Indian Army has also ordered a few units. The missile seeker is being upgraded to track and shoot down drones. The IAF also has 18 Israeli SPYDER-MR systems developed by Rafael Advanced Defence Systems and Israel Aerospace Industries (IAI). The quick-reaction SAM is capable of engaging aircraft, helicopters, unmanned air vehicles, drones, and precision-guided munitions up to range of 35 km.[10] Barak 8 is an Indo-Israeli SAM developed for the Indian Army and Navy. An ER extended range variant with 150 km range is also under development.

The Prithvi Air Defence (PAD) is an anti-ballistic missile developed to intercept incoming ballistic missiles outside the atmosphere (exo-atmospheric), with a maximum interception altitude of 80 km. Advanced Air Defence (AAD) is an anti-ballistic missile designed to intercept incoming ballistic missiles in the endo-atmosphere at an altitude of 30 km. Prithvi Defence Vehicle Mark-II, was used in March 2019 for India's ASAT test. The three stage PDV Mk-II has the capability to shoot down targets moving at 10 km per second in orbits as high as 1,200 km.[11] India has a new missile defence program which will be focused solely on intercepting cruise missiles. The XR-SAM (Extra Long Range Surface to Air Missile) is supposed to be the DRDO's long range (350 km) SAM program which is still evolving.[12] It is clear that India's SAM program is mostly supported by Israel. There are still many development issues that need resolving, and the same need to be given the highest attention. In addition, inductions need to be hastened. Numbers clearly indicate that they are inadequate to take on any Chinese air threat, leave alone a two-front war.

In the meantime, India has ordered five regiments of the S-400 Triumf SAM system from Russia. Each regiment will have two battalions. Each battalion consists of eight launchers and 32 missiles in addition to acquisition and engagement radar systems and a command post.[13] The S-400 uses four missiles of different range capabilities: the very-long-range 40N6 (400 km), the long-range 48N6 (250 km), the medium-range 9M96E2 (120 km) and the short-range 9M96E (40 km). The declared anti-stealth range is 150 km. Maximum targeting range (detection radius is wider) for a ballistic target with low RCS is claimed to be around 230 km. For a target with RCS of 4 square metres is 390 km, and for targeting of strategic-bomber sized types is 400 km. The S-400 is able to intercept cruise missiles at a range of about 40 km due to their low altitude flight paths. It can engage targets at speeds of up to 17,000 kmph (Mach 14), and altitude of 30 km.[14] The system is mobile and can move cross-country and has a reaction time of 9-10 seconds. Key advantages of the system include its ability to track a high number of stealth targets, high modularity, and high mobility. In 2017 the S-400 was described by *The Economist* as "one of the best air-defence systems currently made."[15] According to Siemon Wezeman, Senior Researcher of SIPRI, the S-400 "is among the most advanced air defence systems available."[16] India is planned to receive the systems starting late 2020. China started getting deliveries in 2018.

IAF Fighter Fleet: Depleting Assets

The IAF is already down to 30 fighter squadrons. All attempts are being made to not let this number go down further.[17] Delay in Light Combat Aircraft (LCA) Tejas had forced the IAF to postpone retirement of a few older MiG-21 variants. The MiG-21 Bison-fleet will continue until 2024 with depleting numbers and lower availability of spares. The IAF has ordered 40 LCA Mk1, 20 in Initial Operational Clearance (IOC) have been delivered. Meanwhile, 16 LCA Mk 1 with Final Operational Clearance (FOC) and four two-seat aircraft are under production and delivery. The target LCA production is 12 in 2019-20 which will thereafter increase to 16 aircraft. The IAF is procuring a further 83 single-seat fighters in Mk 1A configuration and 10 twin-seat aircraft. An official order is expected to be placed by end 2020 with induction starting from 2023. The Mk1A will have improvements such as an advanced AESA Radar, an EW suite, a mid-air refuelling probe, and will incorporate weight reduction along with easier service maintainability. The LCA Mk II will now be an aircraft with canards. It is already being called the Medium Weight Fighter (MWF).[18] HAL will be assembling the initial two pre-production aircraft which ADA has an ambitious plan to fly by end of 2023. It will have the more powerful General Electric F-414-GE-INS6 engine. The aircraft will require extensive flight testing. The Series Production aircraft are targeted by 2028. It will eventually replace the Mirage 2000. In the long term, the IAF plans to have 18 squadrons of LCA.

The IAF's dedicated strike aircraft fleet includes Jaguars and these are being modernized. The Mikoyan MiG-29 have also being upgraded with multirole capabilities and now has increased internal and external fuel, aerial refuelling probe, new avionics including Zhuk-M radar, and new air-to-air missiles.[19] Twenty-one additional MiG 29s are being acquired.

The Dassault Mirage 2000 aircraft are being upgraded to Mirage 2000-5 Mk 2 standards with modern avionics including RDY-2 radar, glass cockpit, helmet-mounted sight, EW systems, and advanced MICA missiles.[20] Aircraft life has been increased by 20 years. Upgrades are moving slowly, but all these fleets should upgrade by 2021. Nearly 272 Su-30 MKI air-superiority fighters are on order and 260 have been delivered till date. 12 more will be acquired to make good those lost in accidents. Aircraft will be upgraded to have the BrahMos cruise missiles and nuclear-capable Nirbhay missiles. Initially, 40 aircraft will be modernized and get active electronically scanned array (AESA) radar, more powerful on-board computers and a new electronic warfare (EW) suite. The process has still to start.

IAF Rafale.
Picture Credit: Dassault Aviation by G. Gosset

Thirty-six Rafale aircraft were contracted in 2016. The first five aircraft have arrived in India and formed the squadron end July.[21] All aircraft will be delivered by 2022. Responses for IAF's Request for Information (RFI) for 110 4th-Generation-plus fighters were received in July 2018. In contention are Dassault Rafale, Eurofighter Typhoon, F-16 Block 70 (Now named F-21), F/A-18 E/F, JAS 39 Gripen NG, MiG-35 and SU-35.[22] The Request for Proposal (RFP) is likely to be issued later in 2020. Even if the process is hastened, the earliest these aircraft can induct is 2026. Meanwhile, DRDO and HAL are working on the Advanced Medium Combat Aircraft (AMCA) which is meant to be a fifth generation plane with stealth and super-cruise among others.[23] The aircraft is still in the initial design stage. The IAF has positioned a Project Monitoring Team (PMT). The AMCA requires development of many technologies and they all have to converge on the timeline. The first flight may take place around 2030 and aircraft inducted around 2035 or later.

IAF's Transport and Helicopter Fleets: Upgrades and Replacement

The fleet of nearly 100 Antonov AN-32 medium transport aircraft are being upgraded. The upgrade got delayed because of troubled Ukraine-Russia relations.[24] The upgrade includes a modern weather radar, new GPS, glass cockpit displays, Ground Proximity Warning System (GPWS), and operational life extension from 25 to 40 years. The IAF has 3 squadrons of Dornier Do 228 light transport aircraft. EADS CASA C-295 twin-turboprop tactical military transport aircraft was shortlisted for 56 HS-748 Avro replacement.[25] No contract has been signed yet. Sixteen are to be bought in fly-away condition. Tata Advanced Systems Ltd (TASL) and Airbus Defence and Space are to make 40 of these in India. The aircraft can carry 9.25 tons load (71 troops) vis-a-vis 6.7 tons of the AN 32, and 5 tons of HS-748. If bought, it could one day be a contender to replace the AN-32 also. The IAF has 17 Ilyushin Il-76s (50-tonne load), and 11 Boeing C-17 Globemaster III (70 ton-load) aircraft. It had initially inducted six Lockheed C-130J (one crashed later) for special operations including troop insertion. Seven more were contracted later. Because of disrupted spares lines, serviceability on the AN 32 fleet could be an issue in the long term. The AN-32 was meant to be gradually replaced by the Indo-Russian Multi-Role Transport Aircraft (MTA) project—a twinjet, high-wing, T-tail design. As per the initial plan, the IAF was to acquire 45 of these medium-range aircraft. The project failed to take off because of disagreements between the two sides.[26]

HAL built the light utility helicopters Chetak and Cheetah that are used for training, rescue, and light transport duties including in Siachen and other high-altitude areas. Part of their role is currently being taken over by the HAL Dhruv Helicopter. Dhruv also has a weaponised version, Rudra. It is inducting 60 helicopters initially. Based on the Dhruv platform, HAL is developing the Light Combat Helicopter (LCH) and a Light Utility Helicopter (LUH). The LCH has shaped well and in many ways is better than the Chinese Z-10 attack helicopter.[27] The Russian Ka-226T LUH has been selected to be made in India by a HAL-Kamov and a Joint-Venture. Contract has still to be signed. Mil Mi-17, Mi-17 1V, and Mi-17 V5 are the mainstays of the medium utility helicopter fleet with nearly 240 aircraft.

The IAF also operates three heavy lift Mil Mi-26 helicopters. All the 15 Boeing Chinook helicopters have arrived and the units formed at Chandigarh and Dinjan in the east.[28] Two squadrons of Mil Mi-25/35 attack helicopters operate in support of the Indian Army. 22 AH-64E Apaches have been inducted as a replacement.[29] There will be a squadron each in the western and eastern sectors. Once the Avro replacement is signed for, the IAF will be reasonably well off in transport aircraft and helicopter assets.

AH-64E Apache.
Picture Source: AINonline. Picture Credit: Boeing

Combat Enablers: Woefully Short

Three IL-76 AEW&C platforms with EL/W-2090 Phalcon radar are in service. Two are expected to be ordered shortly. The IAF meanwhile has inducted two Defence Research and Development Organisation (DRDO) Embraer ERJ-145 based AEW&C aircraft, 'Netra'. Initially, three aircraft have been developed, and two inducted into the IAF.[30] India has plans to purchase up to six militarised Airbus A330s for DRDO AWACS. The Defence Acquisition Council has given Acceptance of Necessity (AON) clearance for two initial platforms.[31] The platform will be modified with large fixed radome which will house the DRDO's indigenously developed radar. India's new AWACS platform will be also capable of refuelling other aircraft mid-air in addition to its AEW&CS role. This project could take 6-8 years or more. My assessment is that the IAF requires around 10 Embraer sized AEW&C and 5 larger aircraft. It has six Ilyushin-78 FRA. Physical availability is limited due to technical issues.[32] Meanwhile, the IAF has been in a search for six additional modern FRA since 2006. The first two attempts were aborted due to issues related to Life Cycle Costs and processes. The IAF initially wanted a two-engine aircraft with two-man crew, effectively ruling out the four-engine IL-78. Besides, there are concerns of IL-78 serviceability which CAG had noted was as low as 49 per cent.[33] The contest appears to be between Airbus A-330 MRTT and Boeing KC-46A. The IAF has proposed that the DRDO AWACS should also double as a refueller. Such a variant of the Airbus A330 already exists.

DRDO's AEW&C 'Netra'.
Picture Source: Defence World.Net

Unmanned Aerial Vehicles

The IAF's unmanned aerial vehicle (UAV) fleet comprises of Israeli Aerospace Industries' (IAI) Searcher II and Heron and they are used for reconnaissance and surveillance. The IAI Harpy is an anti-radar combat UAV, and the DRDO Lakshya is used for aerial targeting practice. The DRDO's Rustom II made its maiden flight in July 2018.[34] It is a medium-altitude long-endurance (MALE) (UAV). The Rustom-I is a tactical UAV with endurance of 12 hours. Rustom-H is a larger UAV with flight endurance of over 24 hours, higher range and service ceiling than Rustom-1. Rustom-II is a UCAV based on the Rustom-H model. Induction and operationalisation of these could still take some years. The DRDO's AURA is planned to be a "self-defending high-speed reconnaissance UAV with weapon firing capability".[35] The AURA will cruise at medium altitude and will be capable of carrying two or more guided strike weapons with on-board sensors for targeting and weapon guidance. This is expected around 2028. The AURA will eventually be named 'Ghattak'. India is looking at more sophisticated large footprint systems like RQ-4 Global hawks. The IAF had sent out RFIs to international suppliers for UCAV with low radar cross-section, long range, high service ceiling, and capability to carry precision-guided weapons in an internal weapons bay. Meanwhile, the Pentagon has cleared the sale of 22 Guardian naval surveillance drones to India, but India is in favour of acquiring an armed drone which operates over both land and sea.[36] Initially India was likely to buy the naval surveillance drones from the USA.[37] Later there are plans to procure 30 Predator-B Drones from the US-based General Atomics worth $ 3 billion, 10 each for the three service.

Air and Ground Based Weapons

The Mirage 2000 aircraft used the SPS guided SPICE 2000 glide bombs to hit targets in Balakot.[38] More of these have now been ordered. The SU 30 MKI is being modified to carry SPICE. Precision bombs which are the preferred weapons for strike. The DRDO, in a joint venture with Russia, is developing the air-launched version of the BrahMos cruise missile (300 km range) which has already been test fired.[39] Newer variants with higher speed (Mach 6), better accuracy, longer range (600 km) and top attack variants are under development. The DRDO is also developing the nuclear capable Nirbhay cruise missile with 1,100 km range.[40] The Indigenous Astra BVR (Beyond Visual Range) air-to-air missile firing has been demonstrated on an Su-30 MKI.[41] The Mirage 2000 upgrade inducted MICA missiles. Induction of Rafales will bring the MBDA Meteor BVR and SCALP-EG cruise missiles. There is also a proposal to acquire Air-to-Ground Modular Weapon HAMMER for the Rafale. The well-beyond their extended-life, S-125 Pechora and OSA-AK SAM-8 surface-to-air missiles are being replaced by indigenous Akash medium range systems. The IAF has eight squadrons in service, and seven more have been ordered.[42] Each squadron contains 125 missiles. The SPYDER (Surface-to-air PYthon and DERby) is an Israeli low-level, quick-reaction surface-to-air missile system covering short and medium range, capable of engaging aircraft, helicopters, unmanned air vehicles, drones, and precision-guided munitions.[43] Six SPYDER-MRs along with Python-5 Surface-to-Air Missiles (SAMs) and Derby SAMs are in service. The IAF has also ordered the long range Barak 8 Indian-Israeli LR-SAM. DRDO is likely to develop the Maitri LLQRM (Low Level Quick Reaction Missile) with MBDA, initially for the Indian Navy. In October 2018, India signed a $ 5 billion deal for five units of Russian S-400 Triumf air defence missile system (400 km). All the systems will be delivered by 2025.[44] India is in talks to acquire the National Advanced Surface to Air Missile System-II (NASAMS II) from the USA, which will be used along with indigenous Russian and Israeli systems for air defence of the National Capital Region (NCR) covering all threats from aircraft, missiles and drones.[45]

Net-Centric Warfare

With the secure encrypted Air Force Network (AFNET) operational since 2010, the IAF has greatly enhanced communications and data transfer for the air defence network. The Integrated Air Command and Control system (IACCS) rides on the AFNET.[46] There is a proposal to connect all IAF aircraft with an indigenous secure operational data Link (ODL) which will allow data transfer to the IAACS system and vice-versa, thus completing all elements for net-centric operations. The signing of Communications Compatibility and Security Agreement (COMCASA) with the USA will allow India to procure encrypted communications for US origin aircraft like C-17s and C-130s. The formation of the tri-service Defence Cyber Agency (DCA) will also support further build up cyber capability, both for defensive and offensive operations.

Surveillance Ground Radars

The entire Indian land mass and areas overlooking the borders and sea are covered by a network of long and medium range radars. Low-looking radars are used as gap fillers and for monitoring low-level threats. The Indian DRDO's Electronic & Radar Development Establishment[47] (LRDE) has been successful in developing the INDRA series of radars, the Rajendra fire-control radar for the Akash missile system, the Central Acquisition Radar (CAR), and the Long Range Tracking Radar (LRTR) which has been developed with the assistance of Elta of Israel. In April 2006, 15 Medium Powered Radars (MPR) EL/M-2084 from ELTA Systems were cleared for the IAF. Also cleared were eight indigenous 'Arudhra' MPRS. Field Evaluation Trials (FET) of the Arundhra were completed only in October 2018,[48] and it awaits induction into the IAF.

Indian Defence Satellites

A tri-service Defence Space Agency (DSA), headed by an IAF officer, and jointly operated by all the three services, the civilian Department of Space and the Indian Space Research Organisation (ISRO) was set up in April 2019, to utilize the country's space-based assets more effectively for military purposes.[49] India currently has remote sensing satellites in orbit, some with a spatial resolution of one metre or below which can be also used for military applications. Indian Remote Sensing (IRS) satellites also have dual military use. As of January 2019, 14 were operational and at least four are dedicated military use satellites.[50] All these Indian satellites are placed in polar sun-synchronous orbits and provide data in a variety of spatial, spectral and temporal resolutions. Some have a spatial resolution of 1 metre or less. Radar Imaging Satellite No. 2 (RISAT-2) has a synthetic aperture radar as the primary sensor, giving it day-night, all-weather monitoring capability and resolution of one metre. CARTOSAT-2A is a highly agile dedicated satellite for the Indian Armed Forces. It carries a panchromatic (PAN) camera capable of taking black-and-white pictures in the visible region of electromagnetic spectrum. CARTOSAT-2B has a resolution of 80 centimetres. The GSAT-7 was launched in 2013 for the exclusive use of the Indian Navy to monitor the IOR with a 2,000 nautical mile 'footprint' and provide real-time input capabilities to Indian warships, submarines and maritime aircraft. The GSAT-7A 'Angry Bird' is meant primarily for the IAF with the Indian Army using 30 per cent of the capacity. It will boost the IAF's network-centric operations. HySIS (Hyper Spectral Imaging Satellite) will provide hyper-spectral imaging, including for coastal zones and inland waterways, and will be used by the Indian Navy. Microsat-R launched in January 2019, served as a target for an Indian ASAT test on March 27, 2019. ISRO had launched four new series Risat satellites, and an advanced Cartosat-3 satellite in 2019. ISRO's annual launch target is 33 missions, of which five would be military satellites.[51]

Defence Budget Inadequacy for Modernisation

The 2019-20 defence budget at Rs. 3,01,866 crore ($ 42.7 billion) is 1.44 per cent of the GDP.[52] The capital budget for new acquisitions is Rs. 1,03,380 crore ($ 14.6 billion). The IAF's total budget share is Rs. 68,949 crore ($ 9.75 billion), 23 per cent of the total. The IAF's capital outlay is Rs. 39,347 crore ($ 5.56 billion). The bulk of this will be used for committed liabilities of earlier purchases such as Apache and Chinook helicopters, Rafale, LCA and S-400. The IAF will need out-of-budget funds for some acquisitions.

IAF 2035: Dream or Possibility

The then IAF Chief, Air Chief Marshal B.S. Dhanoa said while speaking at a seminar 'IAF Force Structure 2035' organised by the Centre for Air Power Studies (CAPS) on September 12, 2018 that the IAF needed to upgrade its fighter jet fleet urgently.[53] "We have to match neighbours Pakistan and China to tackle a two-front war," he said. In a conventional battle, the air battle had to be won, he said. China, he said, had some 1,700 fighter jets while Pakistan has some 20 squadrons (16-18 planes in each). The IAF as of now has 30 squadrons. "Even if we get the 42 squadrons as mandated, we will still be behind the collective strength of Pak-China," Dhanoa said. The IAF is deficient of the authorized 42 squadrons by 12. Typically two squadrons cost around US$ 9 billion. By 2035, the IAF will need to make good the deficient 12 squadrons and buy an additional 9 for new phase-outs, thus needing a total acquisition of 21 squadrons. That would cost close to US$ 100 billion (Rs. 7,00,000 crore). Where is that money? Is the IAF's target to make good the 42 fighter squadrons by 2035 a pipe dream? To achieve the target, in 2035 the IAF should have two squadrons each of Mirage-2000 and Jaguars, 14 Su-30 MKI, two of Rafale, 14 of LCA Variants, two AMCA and six of the newly selected fighters, making a total of 42. That would mean building an average of 18 LCA a year. To achieve all this, the defence budget has to be at least 2.5 per cent of the GDP for the next two decades. Realistically, the IAF's capital budget should have been over Rs. 50,000 crore this year. Most analysts feel the IAF requires 50 squadrons for a two-front war. In addition to the combat aircraft, the IAF urgently needs to replace the HS-748 and, in a phased manner the AN-32. Replacement aircraft, C-295, having already been selected, procurement and the 'Make-in-India' by Tata-Airbus joint venture must unfold.[54] Also, there is an urgent need for more AEW&C, FRA, procurement for which has been struck for nearly a decade.

IAF Fighter Fleet Rationalisation and Maintenance

While the trend world-wide is to have limited fighter fleet types, the IAF is mired in a multiplicity of fleets with Jaguars, Mirage 2000s, Su-30s, MiG-21s and 29s, LCA and the soon-to-be-inducted Rafale. It will be a logistics nightmare. Larger single fleets can amortise costs and maintain decent spare backups. Ideally, the IAF could have settled for SU-30 MKI, Rafale and LCA fleets in the long run with new fighters and the AMCA later. The IAF needs to radically cut the diversity of combat aircraft in its inventory, rationalize its force structure and streamline its logistics set-up. The experience with Soviet/Russian fleets has been low serviceability due to long component repair/replacement cycles. In an 800 aircraft inventory, every 5 per cent increase in serviceability would mean adding 40 aircraft, which means adding more than two new squadrons. All fleet maintenance planning is based on government authorised 75 per cent serviceability. In reality, the serviceability is an average 55-60 per cent. By just reaching 75 per cent, the IAF would add nearly 7-8 squadron strength. This needs to be achieved by innovative inventory holdings and spares management.

Need to Consider: Long Range Bombers

China has a sizeable bomber fleet and it is investing in developing a stealth bomber. Both Russia and the USA have always had a strong bomber fleet with large Cruise missile weapon loads. Defence and Security analyst Bharat Karnad suggests that exorbitantly priced aircraft carriers are unaffordable in the age of hypersonic and supersonic missiles, and a military liability. The Indian naval budget should instead prioritize nuclear-powered ballistic missile-firing and attack submarines, and a surface fleet of multipurpose frigates. Similarly, he suggests that the IAF[55] should consider lease or buying 1-2 squadrons of Tu-160M2 'Blackjack' strategic bombers from Russia as the manned, recallable, vector in the country's nuclear triad. That apart, a large bomber will allow a great conventional punch with Cruise missiles, long range AAMs, and capacity for DEWs. The IAF needs to carry out a cost versus operational benefit analysis.

Hypersonic Weapons

China has plans to have a fleet of nearly six aircraft carriers by 2035. These are invariably exorbitantly priced. India will not be able to match the numbers, and at best will have 2-3 carrier groups. In the age of hypersonic and supersonic missiles, some are already considering that the carriers could be made into a liability, as it will be very difficult to intercept a hypersonic weapon. The USA leads research and development in this field, closely followed by China and Russia. Australia, Japan and the European Union also have civilian and military hypersonic research projects. India conducted the maiden launch of a new hypersonic technology demonstrator vehicle (HSTDV) in June 2019 that will have futuristic applications for next generation missiles and aerial systems.[56] The Indo-Russian joint BrahMos-II will be a hypersonic cruise missile currently under development. At Mach 7, it will have double the speed of the current BrahMos missile. Development could take 7–8 years.

China's Rationalised Defence Budget: Lessons for India

China's new defence 'White Paper' published by the State Council of the People's Republic of China in July 2019[57] is an open strategic challenge to the USA, though it does not in any way suggest conflict. It highlights China's defence and security posture. The India-specific content of that document is small. It however publishes the composition of China's military budget between 2010 and 2017, and compares the defence spending patterns with those of India. The White Paper indicates that significant resources have been allotted to the PLA Navy, obviously in line with their intention to become a global power. While China continues to outspend India by nearly four times, it is important to analyse what China spends on and as military spending priorities. In 2010, China's defence budget was a little more than twice that of India. Both militaries continue to spend considerable sums on personnel costs. India's increase on that count has been significantly more, mostly due to high pensions bill.

In consonance with the new military strategy of fighting 'local wars under high-technology conditions', Beijing trimmed its military personnel size by 500,000 soldiers. The size of the Chinese army itself was reduced by almost 19 per cent. The goal is to move away from a manpower-heavy army-centric military. It is, however, the amount that China spends on new defence equipment compared to India that is significant. China's new defence equipment acquisition budget is nearly five times that of India. This is causing a significant gap in both technologies and numbers between the two. While China spends equivalent of 75 per cent of the capital outlay on personnel costs, in India the figures for personnel costs is 2.3 times the capital acquisition budget. This is something that requires urgent attention.

Changing Global Positions: Rising China and Muscle Flexing Russia

The USA has lived a generation after 1991 without competition from Moscow. U.S. national security strategy is looking at the new "contest for power", with the re-emerging Russia and China actively competing for their place on the planet.[58] Inter-state strategic competition has overtaken terrorism as the primary concern for the U.S. national security establishment. The new dynamics of strategic conflict and competition vis-à-vis Russia and China is the issue.[59] Russia's repeated invasions or aggressions against its neighbours since 2008, its reported use of illegal chemical weapons on NATO soil in 2018, the collapse of the Intermediate-range Nuclear Forces (INF) Treaty, Russia's development of a new arsenal of nuclear delivery systems, and the provocative intervention in Syria are issues for the USA. On the other hand, in economic, demographic, and political terms, Russia is clearly a declining power. However its penchant for aggressive and even exuberant geopolitical risk-taking, and its continued possession of military and technological tools has to be factored.[60] The USA has used sanction-tools like the Countering America's Adversaries Through Sanctions Act of 2017 (CAATSA) against Russia.

China's competition presents much more different and difficult challenges. Its size and economic dynamism, and its increasing desire to convert raw demographic weight and its impressive economic rise into geopolitical power is aggressive and visible.[61] China's rise has been driven by export-led growth and has ridden the liberal, open economic order that the USA had created and sustained. China is also a challenge because of the deep economic connections between it and the USA.

Growing China is already acting with increasingly aggressive self-assertion in the world. This self-assertion is not restricted to territorial self-aggrandizement in the South China Sea, but has a clear agenda of dominance in the Indo-Pacific region and on the world stage later.[62] China and Russia have in mind a world of the authoritarian model. China is pursuing what journalist Nicolas Kristof once called "Market Leninism", ensnaring ever-greater portions of the developing world in manipulated debt dependencies and "neo-colonial" economic relationships.[63] Beijing has spent the last quarter-century implementing a strategy dedicated to challenging and undermining U.S. power and influence in the world. Some of the actions by USA include imposing and implementing sanctions interdicting technology-transfer, preventing Chinese technology theft, and multilateral diplomacy mobilizing diplomatic coalitions to support U.S. emerging grand plans. Containing China requires a new strategy with new ideas in accordance with the rapidly changing circumstances. With the global balance of power still evolving, India will have to keep its own strategic interests in mind, and balance the tilt towards the USA with continued engagement with China.

India's Pakistan-Centric to China-Focused Infrastructure

Till very recently, India's major airbases and radars were positioned on the Pakistan border. The bulk of the military assets were looking West. Airfields are high-value targets. Aircraft on the ground are highly vulnerable and require blast protection. Most eastern airbases were bare runways with no blast protection shelters, and at best a few Operational Readiness Platforms (ORP). They were mostly used for fighter training and air maintenance operations. India's emerging economic strength, self-confidence and doctrinal maturity allowed the IAF to look beyond its borders and reach out into the seas beyond Indian island territories. China's focused concentration on building air power also made India rethink its air strategy. Even the political class and strategic think-tank circles realized the importance of air as a decisive instrument of power projection. The lessons learnt during Kargil also helped the IAF understand the real dimensions of fighting an air war at Himalayan heights. The laser weapons employed by the IAF could take out small bunkers, and successful strikes on the Pakistani main administrative and logistics hub at Muntho Dhalo, dealt a major blow, and changed the way it would fight a possible war across the Himalayas. It was clear that the application of airpower has to be precise and proportionate. The IAF's all-new platforms and weapon acquisitions require operational operating altitude clearance of 6 km and higher. It has now deployed front line fighter aircraft of the SU-30 MKI class in the eastern sector. One Rafale squadron is also slated to be deployed in the east. The C-17s have operated from eastern airfields including ALGs. A unit of C-130J is located in the east. A unit each of the new helicopters, Chinook and Apache, will be in the east.

Strategic Effect and Conventional Deterrence

Air power is inherently strategic in nature. Paucity of defence funding has forced India to evolve the doctrine that air power must deliver strategic effects through basic conventional means.[64] Only air power can achieve this. While control of the air remains a desirable state, the IAF has limits to which it can achieve it against the PLAAF. India has also to defend itself against a possible sizeable Chinese surface-to-surface missile (SSM) attack. U.S. think tanks are closely evaluating the IAF doctrine because India is seen as a possible counterbalance against China. Sino-Indian military competition and distrust remain. China continues to create military posts around India. Chinese rapid reaction forces are deployed close to the border. India cannot match China's numerical strength, but the IAF would provide a sufficiently strong "deterrent force".[65] The IAF now has credible "strategic reach", in turn, deterrence. It is looking at reach from the Persian Gulf to the Straits of Malacca, using long range aircraft supported by FRA and AWACS. It is not only the range and weaponry, it is the effect achieved that makes a mission strategic. Targeting of the leadership using air power, as was the case of the U.S. strike against Osama bin Laden, and later by the IAF during the Balakot strikes could actually deliver strategic effects. Strategic air lift is another area of strategic reach and strategic effect. Inter-theatre movement of a large force to support a counter-offensive or reinforce defences can have a strategic effect.

Jointmanship: A Key to Success

Due to the multi-dimensional nature of conflict, increasing levels of synergy amongst the armed forces and civil agencies is operationally critical.[66] The IAF is going to be a key element in support of the surface, maritime and sub-surface wars. It has air assets to support such effort. Lessons from peace time joint exercises and also from past wars is that much more needs to be done on the jointmanship front. The Services continue to train and fight in compartments. Formation of the CDS is likely to support jointmanship. Some level of service upmanship exists. Elements such as Tactical Air Centres (TAC) and Maritime Element of Air Force (MEAF) are in place. However, during exercise planning, the Air Force is often brought in into the discussions when all the planning is already over. There is a need for genuine respect for each other's capability and use it to achieve national objectives. The three Services need to train jointly more regularly. Joint operations in all past wars have paid great dividends, whether it was the Tangail air drop, the Meghna heli-lift, or the Longewala air strikes. During the Kargil conflict, the air attacks on the Pakistani logistics hub at Muntho Dhalo finished Pakistani ability to sustain their campaign. The IAF and Indian Navy need more coordination for maritime strikes. Three new tri-service Special Ops Division and Cyber Agency and Space Agency would increase coordination and operational capability.

Future of Air Dominance

In its quest to dominate the aerial battlefield of the future, the USAF is already looking to replace its traditional fighter jets with a network of integrated systems disaggregated across multiple platforms.[67] They are developing technology to deter current and emerging threats and to produce non-linear, game-changing combat capabilities. It could be a "family of systems" to address the range of threats in a highly contested environment. Both Russia and China continue to close the capability gap with the USA, by building long-range missiles, anti-satellite and anti-aircraft weapons to foil the ability of US forces to penetrate. The new air strategy includes both stand-off capability and penetrating forces, with increased dependence on space and cyber to infiltrate enemy defences and defend own networks.

China could defeat a force of 5th generation fighters by expending enormous numbers of UAVs to force U.S. planes to use all their missiles and thus expose themselves to attack. Advances in communications and sensor integration have made it possible to take advantage of munitions carried by aircraft away from the fray, or referred to as 'Arsenal Planes'.[68] The future is thus in the arsenal plane, hypersonic weapons, directed energy, autonomy, and electronic attack. New combat planes will work to find the most effective combination of speed and manoeuvrability, payload, and range for the platform. The right level of stealth, or low-observability, will also be considered.

The Americans, Chinese and Russians are all working on the concepts of 'Air Dominance' for 2040 and beyond, and evolving the sixth generation fighters which will start inducting around 2028.[69] Fighter Bomber is a platform and therefore still likely to stay. More and more of these will become uninhabited, or optionally manned. There will be dramatic changes in the aerial platform performance and aerial weapons. The clear line defining Atmosphere and Space will become indistinct. Aerospace will soon become a common domain with more aerospace-craft routinely transiting up and down, taking advantage of each. Combat engagements will be at much faster speeds and at much greater distances. The development of a hypersonic aircraft would forever change the ability to respond to conflict.

The one who controls space will control the air in future. Aerospace craft will aim to seize control in establishing dominance/supremacy over the enemy's aerospace assets. They will operate under the control/co-ordination of space-based Early Warning and Control satellites with increased Artificial Intelligence (AI). Satellite/aircraft based kinetic and Directed Energy Weapons (DEW) will soon be a reality and will be used for aerial or surface attack. Even if aerospace supremacy cannot be established, a "degree of dominance" in the air-space bubble in a given area and given time-space without prohibitive interference by opposing air forces will be desired. Since India is still at the stage of mastering 4th generation technologies, some analysts feel that it would be best to skip the fifth generation aircraft stage and start working on sixth-generation technologies. Whether it is physically possible remains to be seen.

Future Conflicts

The uncertainties that will accompany drawdown of US forces from Afghanistan and its influence on regional security dynamics is likely to have a negative impact on the regional security environment. India has to prepare for the geo-political shift to the Indo-Pacific region and the rise of

China as a significant aerospace power. Future conflicts will be short, swift and intense engagements against a nuclear backdrop which may be followed by long stabilization periods. A quantum jump in precision and lethality of weapon systems may result in non-linearity and increased tempo of operations. Often, as in the case of the Balakot strikes, operations could be in an ambiguous zone of 'neither peace nor at war'. Wars will increasingly multi-domain varying from non-contact to contact warfare. The IAF will have to build deterrence and have the ability to dominate the air. It will have to induct modern systems for situational awareness, intelligence and precision strike ability. It will require fixed and rotary wing tactical and strategic air transport assets. Transformation would be spread over a period of time. The IAF is looking at a horizon of 15 years. Technological advancements and evolving changes would have to be factored. Artificial Intelligence (AI) has great applications for air power. India must take an early lead. Future security challenges will be more and more complex, multi-dimensional and non-traditional in both kinetic and non-kinetic form. The IAF would need to think differently to be able to tackle the various asymmetric and non-traditional security threats and would require more innovative, out-of–the-box solutions which would leverage the prevalent technology. Seamless integration of the armed forces and the Ministry of Defence is critical for both capability building and focused operations as per national political directives. It will help better visionary planning and a commitment of national resources. For India to be secure, the IAF must continue to touch the sky with glory.

Critical Technologies for India to Concentrate

Future aerial platforms will feature AI as a decisive aid. They will also have Positioning, Navigation, and Timing (PNT), and communications that allow big data movement between the aircraft. Path of risk reduction by prototyping, technology demonstration, and systems engineering work is possible before creation of an aircraft actually starts. It will not just be an aircraft, but a system of systems including communications, space capabilities, standoff, and stand-in options. The next-generation USAF fighter could be larger and more resembling a bomber. Analysis of thousands of air engagements have found that long-range weapons and sensors have dramatically decreased instances of dog-fighting.[70] So, small size, high speed, and manoeuvrability may be less relevant and easier to intercept. Aircraft will rely on enhanced sensors, signature control, networked situational awareness, and very-long-range weapons to complete engagements before being detected or tracked. Larger planes would have greater range that would enable them to be stationed further from a combat zone, have more powerful radars and greater IR detection capabilities, and carry bigger and longer-range missiles.

Artificial Intelligence (AI), smart structures, and hybrid systems will dictate the future. The future is in self-healing structures, breakthroughs in propulsion, materials, power generation and weapon technology. Self-healing structures in particular would pose a significant advantage over modern-day aircraft, remaining airborne despite taking heavy fire. The new generation of engines will allow ultra-high altitude super-cruise. The avionics are supposed to withstand next generation electronic attack and cyber-attack, have passive detection, and integrated self-protection. Future fighters are expected to use advanced engines such as 'Adaptive Versatile Engine' technology[71] to allow longer ranges and higher performance, where the ratios of bypass and compression airflow can be made variable to improve efficiency. Any system must be designed with the aim to maintain a competitive advantage in an austere budget environment. The Passive Aero-elastic Tailored (PAT), a uniquely designed composite wing, will be lighter, more structurally efficient and have flexibility compared to conventional wings.[72] This wing will maximize structural efficiency, reduce weight and conserve fuel. Hypersonic cruise, fuel cell technologies, hybrid sensors, improved human-machine interface using data analytics and bio-mimicry, combination of materials, apertures and radio frequencies that ensure survival in enemy territory are under development.

Things will be built faster, better and more affordably, using 3D printing yet ensuring quality and safety standards. Additive 3D manufacture creates a world with spare parts on demand, faster maintenance and repairs, more effective electronics, and customized weapons.[73] Nano-materials will control sizes, shapes and compositions, and significantly reduce weight, yet create stronger structures for air and spacecraft, and drive down costs.

Uninhabited Aerial Systems and Swarming Technologies

Uninhabited aircraft technologies are already proven, and the future is Uninhabited Aerial Systems (UAS). The world is in transition. Solar-powered UAS will have long endurance. Optionally manned aircraft may become the norm. The USA and China are working on hypersonic (Mach 6) Strike Bombers which are likely to be optionally manned. The US Army's dramatic shift to a nearly all-unmanned flight over the next three decades is embedded in the UAS roadmap. The USAF's UAS vision document indicates that by 2047, every mission would be unmanned.[74] China has invested heavily in unmanned systems. They are already the world leaders in small UAVs and have demonstrated swarming. India's indigenous UAV and swarming, and counter swarming plan, is currently inadequate and must unfold very quickly.

Evolving Future Weaponry

Scramjet based missiles are already a reality. Hypersonic Cruise missiles and glide vehicles are already being inducted by major air forces. Continued experiments with DEW and lasers, used for defensive as well as offensive measures, delivering effects at the speed of light, will shape future warfare. New aircraft will be as much about reusable weaponry (lasers) as it is about expendable weaponry. New air-to-air missiles, could also be powered by improved solid rocket motors having synergized control enabled by combined aero, attitude control and thrust vectoring.[75] The missiles will have improved 'high off bore sight' for rear hemisphere kills and 'lower cost per kill.' The missiles would also incorporate energy optimizing guidance, navigation and control. Range would be a big factor to counter potential adversaries with Chinese PL-15. They will have multiband, broad spectrum to aid survivability and reaching the target. No aircraft is invisible, and using standoff weaponry early in an air campaign to open up weaknesses in an enemy's air defences will be critical. Development of solid-state airborne laser capability is already underway. The solid-state laser systems defensively create a sanitized sphere of safety around the aircraft, shooting down or critically damaging incoming missiles and approaching aircraft with laser turrets, and even attacking targets on the ground with pinpoint precision or shooting down ballistic missiles and other traditional targets.[76] Controlling the aircraft's heat signature while using laser weaponry would have to be managed. Laser weapons are expected to be mounted on next-generation air dominance fighters by the 2030s.

Data Handling and Network Approach

The demand for streaming high-quality data requires bandwidth, which involves innovating sensor/processing systems. Mission computer systems and network-centric payload processing units enable onboard data fusion prior to sending to digital links. Thermally efficient, high-performance computing onboard the aircraft is essential. Next-generation avionics would be smaller, more efficient and capable of operating under extreme conditions. Gallium Nitride (GaN) is a semiconductor material that is more efficient, easier to cool, and improves reliability for radars.[77]

Effect-Based Operations to Information-Centric Operations

Airpower strategy for the 21st century is shifting from Effect-Based Operations (EBO) to Information-Centric Operations (ICO).[78] Future wars are likely to be won by the side that can most effectively dominate the informational realm. Air, space, and cyberspace forces have a unique capability to collect information in multiple domains, rapidly fuse, analyze, and disseminate that information and allow an effect to be delivered. To successfully achieve EBO, information gathering will be crucial to effectively fight and win future conflicts. Developing and delivering air superiority for the highly contested environment in 2030 requires a multi-domain focus on capabilities and capacity. Technologies such as hypersonic weapons and swarming tactics need a unifying idea for employment. Such innovations must be paired with valid concepts of operation to make them effective in the expected operational environment. Critical to these emerging technologies is the voracious need for information, and more accurate, timely and actionable intelligence. Hypersonic weapons and swarms of mini-drones are only effective if the shooter knows where the target and its defences are. Cyber weapons are only effective if one understands an adversary's network(s). These types of capabilities require decision-quality intelligence, which is derived from the rapid collection, fusion, analysis, and delivery of information. Importance of Information Dominance provides a strategy for integrating emerging intelligence, C2, and weapons capabilities, and explains how each combat capability in the Air Force is part of a larger process that collects information in all war-fighting domains, considering the challenges of balancing capabilities with modernisation priorities and devising a necessary acquisition strategy in the light of such challenges as maintaining critical skills and addressing budgetary shortfalls and operating within a shrinking R&D community and technological/industrial base. Air, Space, and Cyberspace Power are the real domains of action in the 21st century.

Critical Aerospace Areas for India to Address

India has to push for indigenisation of all military related technologies and also build up its domestic production base including engaging the private sector. This has to be done as a national mission, bypassing bureaucratic hurdles. The emphasis has to be on long-range offensive precision strikes in air and space with enablers like satellites, FRA and AWACS, and also to have the ability to counter any aggression. Space has become an enabler for all strategic domains and military arms. India has both scientific expertise and ability in space technologies. This needs to be harnessed for military power and economic growth. China is way ahead of India with its large constellation of satellite navigation systems, and also in space based C4ISR. Luckily, India is moving steadily in that direction.

India has also to build an arsenal of short range conventionally armed ballistic missiles and Cruise missiles with ranges up to 1,000 km to take-on targets like airfields, logistics nodes and ammunition dumps in the Tibet region. The IAF has to increase focus on AI based systems and electronic warfare. India must harness its cyber and information warfare resources. It has to have multi-layered air defence systems that can cover threats emanating from across the Himalayan mountain ranges.

Operationally, the IAF has to use innovative tactics and top end fighters to prevent the PLAAF from achieving any degree of dominance even in a limited part of the theatre of operation. The IAF has to contend with PLAAF H-6K bombers armed with six DH-10 Cruise missiles, which have a range of around 1,500 km. These bombers have a combat radius of 1,800 km, which means that they do not need to get airborne from airfields in Tibet and can launch their Cruise missiles on critical Indian military targets from well outside any kind of air defence umbrella that the IAF can put in place over the next decade.[79] It will be important to operationalise the Rafale quickly. Also, 36 is too small a number. At least 36 more are needed to be bought to make a reasonable deterrent against China. The PLAAF may have limitations in terms of fighter sortie generation rate in the TAR. But it has been training with the PAF regularly and has a good sense of tactics followed in the sub-continent. The IAF has to be conscious of this. Since India is normally not likely to begin a conflict, the IAF must have the capability to withstand any initial PLARF strike and be able to launch a punitive return strike.[80] In view of repeat PLARF SSM strikes, the IAF's sortie generation is also likely to be affective. It will also have to factor Chinese SAMs that will restrict IAF operations in the TAR or cause attrition.

Any operation will require total jointness and synergy between the political leadership and all the elements of state power including the three armed forces. The IAF's recent emphasis on parallel operations and continuous integration with the Indian Army in high-altitude terrain will result in immediate impact in and around the TBA.[81]

Choices of Coalition

The USA has been trying to urge like-minded states to take a stand against the Chinese position in the South China Sea, and to assert their freedom of navigation despite Chinese maritime claims. More and more nations of the region have begun to assert their rights in the South China Sea. The Quadrilateral Security Dialogue (QSD, also known as the Quad) is an informal strategic dialogue between the USA, Japan, Australia and India, initiated in 2007 by Prime Minister Shinzo Abe of Japan. The dialogue has been followed by a joint military exercise titled Exercise Malabar. India has been maintaining good relations with most ASEAN nations and others in the Indo-Pacific nations. It also has close cultural links with many. Even if they will not side with India, they will not work against India. It is in the best interest to keep engaging Quad in military exercises. India has to keep the pipeline open for top of the line defence systems from both the USA and Russia, till indigenous systems are in place.

Way Ahead for India

The IAF needs to address issues of protection, redundancy of capabilities, and survivability of assets. It needs to work jointly with the other Services, increase flexibility, interoperability, and sustainability; and preserve India's ability to project air power into potentially contested regions. It must identify robust long-range strike capabilities as essential to deterrence and for providing a spectrum of response options. This could translate into establishment of an integrated strike force, leveraging the capabilities of next-generation bombers, missiles, and UCAVs, in particular, stealth long-range aircraft. The IAF needs to augment and replace its aging FRA fleet.

The IAF must increasingly become empowered to wage asymmetric warfare, kinetic and non-kinetic attacks such as cyber war, jamming, anti-satellite (ASAT), and electromagnetic pulse (EMP). Improving joint Service integration, planning, and inter-operability is increasingly important. India must develop long-range, offensive systems such as low-observable manned and remotely piloted strike aircraft, precision missiles, and intelligence, surveillance, and reconnaissance (ISR) platforms to penetrate heavily defended Anti Access/Area Denial (A2/AD) environments. India must develop next generation space systems and develop rapid replenishment capacity, including micro-satellite technologies and systems and new launch capabilities. The survivability and resilience of India's space networks are imperative. Cyber-attacks on military networks have shown that critical systems are vulnerable. The IAF's cyber operations must be capable of operating in a contested cyber domain to support vital land, sea, air, and space missions. Priorities must include developing capabilities to protect essential military cyber systems and to speed their recovery if an attack does occur. C2 must be resilient and dependable even in a post cyber-attack degraded environment. India has to have the means to protect its systems from a possible Chinese high-altitude nuclear detonation that can produce EMP that can affect electronics-dependent national infrastructure.

The IAF needs to adapt to the demands of the information age. There is an increasing demand for ISR capabilities able to access and persist in contested airspace in order to track a range of high value mobile and hard-to-find targets, such as missile launchers and underground bunkers. This increases the need for stealthy, survivable systems and the development of next-generation unmanned platforms. Attention is needed for what is called a "family of systems" that could be deployed in multiple ways with maximum versatility depending on requirements. The IAF needs to develop the next generation of long-range strike platforms. There has to be a greater push for unmanned or optionally-manned platforms. India needs to quickly complete its own secure satellite navigations systems. In any case, the IAF has to look at alternatives to excessive reliance on GPS or other space based systems in case of denial of the same through jamming. DEW and laser weapons is where the future is. It is time to invest in these technologies in a big way. India has to build a capability to combat anti-satellite threats, advanced surface-to-surface missiles, and new-generation ballistic and cruise missiles.

Enhanced service jointness entails assembling the appropriate force mix for a contingency. Inter-dependence is closely related to the concept of jointness. Minimizing duplication of effort by the Services should be done because it creates inefficiencies, promotes redundant concepts of operations, and causes cross-service frictions. Without inter-dependence, Service jointness will not reach its full potential. There is a need for joint force planning to reach a level of force inter-dependence that leverages capacities across the Services without duplicating efforts. For example, the IAF's manned aircraft and the Indian Navy's maritime patrol and UAVs could be integrated into a joint long-range maritime strike mission.

Irregular warfare demands special operations forces (SOF) like the IAF's Garud with high mobility, and regular training cross-training. India's tri-service Special Operations Division should ensure that. Air mobility and ISR are key enablers for SOF operations, and the C-130 and the IAF's helicopter fleet are great assets. India needs to build capability for out-of-area operation in the Indian Ocean littorals including the defence of India's island territories.

India has to make its armed forces lean and mean. A starting point for that will be acquisition of modern platforms and force multipliers. India's personnel costs form a high part of the defence budget. For increased capital acquisition budget share, either the total defence budget must go up or personnel cost reduced. With the current threat perception, the numbers cannot be pruned much. The service chiefs must sit with the government to find a good solution for better force readiness, modernization, and even day-to-day operations. In any case, the MoD, IAF, DRDO, DPSUs and industry must coordinate R&D and systems production efforts more closely.

Like the Indian Navy and many other major air forces, the IAF must reclaim and get more directly involved to push research and development and aircraft building. Catch up on stealth technologies in the face of ever more sophisticated enemy integrated air defence environments. Push fifth-generation fighter aircraft. Develop new technologies in electronic warfare, which continues to be a key enabling capability for conflict operations across the warfare spectrum. Continue development of unmanned technologies for missions and programs focused on increased loiter time, and enhanced munitions capabilities. The IAF must support innovation. Early warning capabilities rely on Air Force satellites and radar networks. India needs to invest in a few dedicated electronic warfare aircraft of the EA-6B Prowler class deployable from land bases and aircraft carriers to monitor the electromagnetic spectrum and actively deny an adversary the use of radar and communications.

Like China's quest for improved aerospace power which is based on the Chinese Communist Party's (CCP) evolving strategic needs, India needs such strategic discussions at the political level. The IAF leadership must re-conceptualize its vision, institutional identity, and force posture to align as closely as possible with future national security settings. It is time to take a call for an appropriate balance between high and low end air and space capabilities based on emerging threats and fiscal constraints. India should consider building "preventive partnerships" to avert conflicts before they erupt. The USA is the obvious contender. Inter-operability with USAF, post COMCASA, and LEMOA should help. As China becomes more independent of Russia, weaning away Russia from China could be the foreign policy initiative for most of the world, especially the USA, Europe and India.

Concluding Remarks

In recent years, India's ties with China have been characterized by dissonance, with China opposing India's entry into the Nuclear Suppliers' Group (NSG), Indian worries about trade imbalances, Chinese hostility towards the Dalai Lama's travels to Arunachal Pradesh and finally India's decision to oppose China's Belt and Road Initiative (BRI), made in no small part due to BRI's flagship project being the China-Pakistan Economic Corridor (CPEC), which impinges on India's sovereignty.[82] Throughout this period, the LAC has seen numerous Chinese transgressions in certain sectors as well as airspace violations. With CPEC, China has inserted itself into the Kashmir dispute. The peculiar bilateral ethos of Sino-Pak relations has created fears in New Delhi about the possibility of a collusive two-front threat. India's polity has therefore accepted the Indian military's need to be prepared for a two-front conflict and the need to build the necessary wherewithal to be able to hold-off a Chinese attack while being engaged in a short punitive war against Pakistan.

Chris Dougherty, of Centre for New American Security, writes in 'Force Development Options for India by 2030', that China's military challenges require the Indian armed forces to update or replace antiquated military strategies, operational concepts, organizational constructs, and equipment.[83] India also faces ongoing terrorist and irregular threats, as well as conventional and nuclear threats from less-formidable regional opponents. Exploit the late-mover advantage to invest in relatively inexpensive offensive capabilities in space, cyber, precision fires, and undersea domains, given India's limited defence budget and pressing strategic demands. India should seek international partnerships in areas of vulnerability. India should accelerate and increase its investments in extending the range of the BrahMos and developing the longer-range

Nirbhay Cruise missile. While India's Navy will likely attempt to keep aircraft carriers as the centrepiece of its fleet, in view of the challenges posed by China, India should focus more on undersea and an air-centric combat force. The Indian Navy may see greater strategic benefits from placing scarce assets in land-based aircraft, missiles, and submarines. He further suggests the need to consolidate procurement, particularly in combat aircraft, to create greater economies of scale and efficiencies in operation, by rationalising fighter fleets. India could consider skipping fifth-generation aircraft and wait to see what technologies emerge in the next generation. Given the time it has taken India to procure fourth-generation aircraft, any decision today by India to procure fifth-generation aircraft is likely to be overcome by events before the aircraft are delivered. He also feels, and I think rightly so, that India's complex bureaucracy-military relations have hampered India's ability to develop and implement effective strategies. India lacks a strong defence bureaucracy or a joint military body to oversee strategy. India could establish a standing joint body of experienced civilian and military strategists and analysts to oversee the formulation and implementation of defence strategy, operational plans and concepts, and force development.

REFERENCES

1. Lao Tzu, Quotable Quotes, https://www.goodreads.com/quotes/50543-there-is-no-greater-danger-than-underestimating-your-opponent
2. Elan Chan, 'Made in China 2025': is Beijing's plan for hi-tech dominance as big a threat as the West thinks it is?, *South China Morning Post*, September 10, 2018, https://www.scmp.com/business/china-business/article/2163601/made-china-2025-beijings-plan-hi-tech-dominance-big-threat
3. Anthony H. Cordesman & Arleigh A. Burke Chair in Strategy acordesman@gmail.com With the assistance of Max Molot, China and the U.S.: Cooperation, Competition and/or Conflict An Experimental Assessment, CSIS, Revised October 01, 2019, p. 194, https://csis-prod.s3.amazonaws.com/s3fs-public/publication/191001_China_Grand_Strategy.pdf
4. Rand.org, An Interactive Look at the U.S.-China Military Scorecard, https://www.rand.org/paf/projects/us-china-scorecard.html
5. Anthony H. Cordesman with the assistance of Joseph Kendall, The PLA Rocket Force: Evolving Beyond the Second Artillery Corps (SAC) and Nuclear Dimension, Centre for Strategic and International Studies (CSIS). October 13, 2016, https://csis-prod.s3.amazonaws.com/s3fs-public/publication/161013_China_Missile_Forces_AHC.pdf
6. Ibid.
7. Anthony H. Cordesman, Ashley Hess & Nicholas S. Yarosh, Chinese Military Modernization and Force Development: A Western Perspective, CSIS, p. 219.

8. Vivek Raghuvanshi, Indian Excludes Foreign Vendors From Its Air-Defence Upgrade, *Defence News*, June 13, 2016, https://www.defensenews.com/global/asia-pacific/2016/06/13/indian-excludes-foreign-vendors-from-its-air-defense-upgrade/
9. Akash Surface-to-Air Missile (SAM) System, Air Force Technology, https://www.airforce-technology.com/projects/akash-surface-to-air-missile-system/
10. Harsh V. Pant & Ambuj Sahu, Israel's arms sales to India: Bedrock of a strategic partnership, orfonline.org. September 04, 2019, https://www.orfonline.org/research/israels-arms-sales-to-india-bedrock-of-a-strategic-partnership-55101/
11. Ashley J. Tellis, India's ASAT Test: An Incomplete Success, Carnegie Endowment for International Peace, April 15, 2019, https://carnegieendowment.org/2019/04/15/india-s-asat-test-incomplete-success-pub-78884
12. U. Sudhakar Reddy, DRDO reveals 350km XRSAM missile details, *Times of India*, December 28, 2019, https://timesofindia.indiatimes.com/city/hyderabad/drdo-reveals-350km-xrsam-missile-details/articleshow/73001348.cms
13. Franz-Stefan Gady, India and Russia Ink S-400 Missile Air Defence System Deal, *The Diplomat*, October 20, 2016, https://thediplomat.com/2016/10/india-and-russia-ink-s-400-missile-air-defense-system-deal/
14. S-400 missile system, Wikipedia, https://en.wikipedia.org/wiki/S-400_missile_system
15. Turkey and Russia cosy up over missiles, *The Economist*, May 04, 2017, https://www.economist.com/europe/2017/05/04/turkey-and-russia-cosy-up-over-missiles
16. Yarno Ritzen, Why do countries want to buy the Russian S-400? aljazeera.com. October 08, 2018, https://www.aljazeera.com/indepth/features/countries-buy-controversial-russian-400-181007205808578.html
17. Ajai Shukla, Air Force chief outlines plan to solve shortage of fighter squadrons, *Business Standard*, October 05, 2019, https://www.business-standard.com/article/defence/air-force-chief-outlines-plan-to-solve-shortage-of-fighter-squadrons-119100401549_1.html
18. Yusuf T. Unjhawala, Indian Air Force should drop its plan to make more Tejas Mark-2s & focus on AMCA fighter jets, *The Print*, September 19, 2019, https://theprint.in/opinion/indian-air-force-should-drop-its-plan-to-make-more-tejas-mark-2s-focus-on-amca-fighter-jets/293883/
19. Lt Gen V.K. Saxena, IAF to buy Upgraded MiG-29s from Russia, India Strategic. July 2019, https://www.indiastrategic.in/2019/07/10/iaf-to-buy-upgraded-mig-29s-from-russia/
20. Mirage 2000 fighter aircraft to be upgraded with multi-function displays, *Economic Times*, (PTI) July 14, 2018, https://economictimes.indiatimes.com/news/defence/mirage-2000-fighter-aircraft-to-be-upgraded-with-multi-function-displays/articleshow/48732142.cms?from=mdr
21. https://www.thedrive.com/the-war-zone/35236/indias-first-french-built-rafale-fighters-have-finally-arrived)
22. Snehesh Alex Phillip, India will purchase 110 new fighter jets in staggered manner, says CDS Gen Bipin Rawat, *The Print*, February 17, 2020, https://theprint.in/defence/india-will-purchase-110-new-fighter-jets-in-staggered-manner-says-cds-gen-bipin-rawat/366798/
23. Anantha Krishnan M., AMCA could fly undetected during dangerous missions, english.manoramaonline, February 05, 2020, https://english.manoramaonline.com/news/nation/2020/02/05/amca-could-fly-undetected-during-dangerous-missions.html

24. Amrita Nayak Dutta, There are 46 upgraded AN-32 aircraft, but none deployed in the Northeast, *The Print*, June 05, 2019, https://theprint.in/defence/there-are-46-upgraded-an-32-aircraft-but-none-deployed-in-the-northeast/245772/
25. Jimmy Bhatia, C-295 Project Ready to Roll Out, *India Strategic*, March 2019, https://www.indiastrategic.in/2019/03/26/c-295-project-ready-to-roll-out/
26. Ilyushin Il-276, Wikipedia, https://en.wikipedia.org/wiki/Ilyushin_Il-276
27. Lt Gen B.S. Pawar, Comparison: India's LCH, Chinese Z-10, Turkey's T-129 Attack Helicopters, October 11, 2019, bharatshakti.in, https://bharatshakti.in/comparison-indias-lch-chinese-z-10-turkeys-t-129-attack-helicopters/
28. https://www.financialexpress.com/defence/indian-air-force-gets-15-deadly-and-heavyweight-chinook-helicopters-to-help-operations-in-the-northeast-region/1972582/
29. https://www.financialexpress.com/defence/apache-firepower-for-iaf-boeing-confirms-completing-delivery-of-worlds-best-combat-helicopters-to-india/2019809/#:~:text=at%20closed%20border-,Apache%20firepower%20for%20IAF!,Indian%20Air%20Force%20(IAF).
30. Rahul Bedi, Indian Air Force receives second indigenous AEW&C system, *Jane's Defence Weekly*, September 13, 2019, https://www.janes.com/article/91161/indian-air-force-receives-second-indigenous-aew-c-system
31. Arun Mathew, India's Defence Acquisition Council Approves Procurement Worth $3.2 Billion, *Defence Post*, November 28, 2019, https://defpost.com/indias-defence-acquisition-council-approves-procurement-worth-3-2-billion/
32. Gp Capt A.K. Sachdev, Flight Refuellers for the IAF: Lessons from Gagan Shakti 2018, *India Defence Review*, Vol. 33.4 Oct-Dec 2018, http://www.indiandefencereview.com/spotlights/flight-refuellers-for-the-iaf-lessons-from-gagan-shakti-2018/
33. IAF hunts for aerial refuellers, again: A saga of cancelled deals, The Week, Web Desk, October 04, 2019, https://www.theweek.in/news/india/2019/10/04/hunt-for-aerial-refuellers-another-never-ending-saga-for-iaf.html
34. Chethan Kumar, DRDO's combat drone Rustom-2 flies for the first time, *Economic Times*, July 14, 2018, https://economictimes.indiatimes.com/news/defence/drdos-combat-drone-rustom-2-flies-for-the-first-time/articleshow/55451267.cms
35. AURA UCAV:: The Future of Indian Airforce, Defence Updates, https://defenceupdate.in/aura-ucav-the-future-of-indian-airforce/
36. US approves sale of 22 Guardian drones to India: Sources, *Economic Times*, (PTI) July 15, 2018, https://economictimes.indiatimes.com/news/defence/us-approves-sale-of-22-guardian-drones-to-india-sources/articleshow/59274184.cms?from=mdr
37. Manu Pubby, India may go for only naval UAVs from US, *Economic Times*, Aug 29, 2019, https://economictimes.indiatimes.com/news/defence/india-may-go-for-only-naval-uavs-from-us/articleshow/70882886.cms?from=mdr
38. Md. Waquar Haider, Everything You Need to Know About The SPICE-2000 Smart Bombs Used by The IAF in Balakot Strikes, *News 18*, March 06, 2019, https://www.news18.com/news/tech/everything-you-need-to-know-about-the-spice-2000-smart-bombs-used-by-the-iaf-in-balakot-strikes-watch-video-2057655.html
39. BrahMos Aerospace, Official Website, http://www.brahmos.com/content.php?id=19
40. Debjit Sinha, What is Nirbhay missile? All you want to know about India's indigenous 1,000-km nuclear-capable missile, *Financial Express*, April 16, 2019, https://www.financialexpress.com/defence/what-is-nirbhay-missile-test-2019-nuclear-range-of-india-pakistan-china-drdo/1549892/

41. Air-to-air missile Astra test-fired from Sukhoi-30 MKI, *Economic Times*, (PTI) September 17, 2019, https://economictimes.indiatimes.com/news/defence/india-successfully-test-fires-air-to-air-missile-astra/articleshow/71167756.cms?from=mdr
42. Initial problems behind us, we have ordered more Akash missiles systems: Top IAF officials, *Business Standard*, (ANI), December 4, 2019, https://www.business-standard.com/article/news-ani/initial-problems-behind-us-we-have-ordered-more-akash-missiles-systems-top-iaf-officials-119120401010_1.html
43. Fayiq Wani, India successfully test-fires Israeli SPYDER missile from Pokhran Test Range, *News Nation*, August 09, 2019, https://www.newsnation.in/india/news/india-successfully-test-fires-israeli-spyder-missile-from-pokhran-test-range-233586.html
44. Russia starts making S-400 missiles for India, to be delivered by 2025, *Business Standard*, (PTI) January 17, 2020, https://www.business-standard.com/article/pti-stories/russia-starts-production-of-s-400-missile-systems-for-india-120011701027_1.html
45. Snehesh Alex Philip, New Delhi to get Washington-type missile shield NASAMS II, *The Print*, February 11, 2020, https://theprint.in/defence/new-delhi-to-get-washington-type-missile-shield-nasams-ii/363077/
46. AFNET to Herald Network Centric Operations in IAF, Press Information Bureau. September 08, 2010, https://pib.gov.in/newsite/PrintRelease.aspx?relid=65621
47. Electronic & Radar Development Establishment (LRDE), DRDO Website, https://www.drdo.gov.in/labs-and-establishments/electronics-radar-development-establishment-lrde
48. Palash Choudhari, Varun Karthikeyan & Anoop Madhavan, DRDO/BEL ARUNDHRA: Medium Power Radar - The Sacred wave, Full Afterburner, April 01, 2018. http://fullafterburner.weebly.com/next-gen-weapons/drdobel-arudhra-medium-power-radar-the-sacred-wave
49. Vivek Raghuvanshi, India to launch a defencese-based space research agency, *Defence News*, June 12, 2019, https://www.defensenews.com/space/2019/06/12/india-to-launch-a-defense-based-space-research-agency/
50. Wikipedia, Integrated Space Cell, https://en.wikipedia.org/wiki/Integrated_Space_Cell
51. Surendra Singh, Isro to launch 5 military satellites this year to boost 'strategic assets in space. *The Times of India*, April 04, 2019, https://timesofindia.indiatimes.com/india/isro-to-launch-5-military-satellites-this-year-to-boost-strategic-assets-in-space/articleshow/68713168.cms
52. Laxman Kumar Behera, Interim Defence Budget 2019-20, Manohar Parrikar Institute for Defence Studies and Analyses, New Delhi. February 04, 2019, https://idsa.in/issuebrief/interim-defence-budget-2019-20-lkbehera-040219
53. Ajay Banerjee, Need to upgrade its fleet urgently, says IAF chief Dhanoa. *Tribune News Service*, September 12, 2018, https://www.tribuneindia.com/news/archive/nation/need-to-upgrade-its-fleet-urgently-says-iaf-chief-dhanoa-651704
54. Dinkar Peri, C-295 transport plane clears negotiations, *The Hindu*, November 12, 2019, https://www.thehindu.com/news/national/c-295-transport-plane-clears-cost-negotiations/article29955949.ece
55. Bharat Karnad, Time for Disruptive Foreign and National Security Policies, Centre for Policy Research, June 06, 2019, https://cprindia.org/news/7838
56. Elizabeth Roche, India successfully test fires hypersonic cruise missile, *Live Mint*, June 12, 2019, https://www.livemint.com/science/news/drdo-test-fires-hypersonic-technology-demonstrator-vehicle-1560353389161.html

57. Anthony H. Cordesman, China's New 2019 Defence White Paper, Centre for Strategic and International Studies, July 24, 2019, https://www.csis.org/analysis/chinas-new-2019-defense-white-paper
58. Dr. Christopher Ashley Ford, Re-learning a Competitive Mindset in Great-Power Competition, U.S. Department of State, March 14, 2019, https://www.state.gov/re-learning-a-competitive-mindset-in-great-power-competition/
59. Karen A. Miller, Rafael Loss, Mike Markey & Mona Dreicer, Strategic Weapons in the 21st Century - The New Dynamics of Strategic Conflict and Competition, Co-hosted by Lawrence Livermore National Laboratory and Los Alamos National Laboratory, March 14, 2019, https://cgsr.llnl.gov/content/assets/docs/SW21-2019-Summary-Final.pdf
60. *Reuters*/Mike Segar, Geopolitical Power Shifts, Report, World Economic Forum, https://reports.weforum.org/global-risks-2018/geopolitical-powershift/
61. Jonathan Woetzel, Jeongmin Seong, Nick Leung, Joe Ngai, James Manyika, Anu Madgavkar, Susan Lund & Andrey Mironenko, China and the world: Inside the dynamics of a changing relationship McKinsey Global Institute (MGI) report, July 2019, https://www.mckinsey.com/~/media/mckinsey/featured%20insights/china/china%20and%20the%20world%20inside%20the%20dynamics%20of%20a%20changing%20relationship/mgi-china-and-the-world-full-report-june-2019-vf.ashx
62. S.D. Pradhan, Expanding Chinese aggressive activities from South China Sea to Indian Ocean: containment a necessity, Chanakya Code, *Times of India*, December 18, 2019, https://timesofindia.indiatimes.com/blogs/ChanakyaCode/expanding-chinese-aggressive-activities-from-south-china-sea-to-indian-ocean-containment-a-necessity/
63. Simon Lester, Is the Trump Administration Pushing for a Cold War with China? CATO Institute, June 10, 2019, https://www.cato.org/blog/trump-administration-pushing-cold-war-china
64. Christina Goulter & Harsh V. Pant, Realignment and Indian Airpower Doctrine, Air University, January 02, 2020, https://www.airuniversity.af.edu/Wild-Blue-Yonder/Article-Display/Article/2007488/realignment-and-indian-airpower-doctrine/
65. Ibid.
66. Joint Doctrine Indian Armed Forces, Headquarters Integrated Defence Staff Ministry of Defence, April. 2017, https://www.ids.nic.in/IDSAdmin/upload_images/doctrine/JointDoctrineIndianArmedForces2017.pdf
67. Lara Selugman, Beyond the Fighter Jet: The Air Force of 2030, *Defencese News*, April 08, 2016, https://www.defensenews.com/air/2016/04/08/beyond-the-fighter-jet-the-air-force-of-2030/
68. Robert Farley, F-35s and F-22s Might Be Obsolete. The Air Force's Solution: 'Arsenal Planes'? *The National Interest*, November 26, 2019, https://nationalinterest.org/blog/buzz/f-35s-and-f-22s-might-be-obsolete-air-forces-solution-arsenal-planes-99882
69. Valerie Insinna, The US Air Force's radical plan for a future fighter could field a jet in 5 years, The *Defence News*, September 16, 2019, https://www.defensenews.com/digital-show-dailies/2019/09/16/the-us-air-forces-radical-plan-for-a-future-fighter-could-field-a-jet-in-5-years/
70. John Stillion, Trends in air-to-air combat implications for future air superiority, Centre for Strategic and Budgetary Assessments. p. 24, 2015, https://csbaonline.org/uploads/documents/Air-to-Air-Report-.pdf
71. Stephen Trimble, USAF starts work on defining adaptive engine for future fighter, *Flight Global*, July 10, 2018, https://www.flightglobal.com/fixed-wing/usaf-starts-work-on-defining-adaptive-engine-for-future-fighter/128748.article

72. NASA, Experimental Wing Proving New Design Methods, September 26, 2018, https://www.nasa.gov/centers/armstrong/features/pat_wing.html
73. Richard D. Aveni, The 3-D Printing Revolution, *Harvard Business Review*, May 2015 Issue, https://hbr.org/2015/05/the-3-d-printing-revolution
74. Colonel Eric Mathewson, AF UAS Task Force, AF UAS Flight Plan 2009-2047, USAF airforcemag.com p. 15, https://www.airforcemag.com/PDF/DocumentFile/Documents/2009/UAS-2047_briefing_Mathewson_072309.pdf
75. Leigh Giangreco, USAF developing next generation air dominance missile, *Flight Global*, May 31, 2017, https://www.flightglobal.com/usaf-developing-next-generation-air-dominance-missile/124213.article
76. Tyler Rogoway, This Is Northrop Grumman's Idea Of A Sixth-Generation Fighter, But Is It Feasible? Jalopnik.com. December 12, 2015, https://foxtrotalpha.jalopnik.com/this-is-northrop-grummans-idea-of-a-sixth-generation-fi-1747680825
77. How Gallium Nitride Enables Smaller, More Efficient Power Supplies The American Society of Mechanical Engineers, March 11, 2020, https://www.asme.org/topics-resources/content/how-gallium-nitride-enables-smaller-more-efficient-power-supplies
78. Tyson Wetzel, Airpower Strategy for the 21st Century, realcleardefense.com, October 16, 2017, https://www.realcleardefense.com/articles/2017/10/16/airpower_strategy_for_the_21st_century_112473.html
79. Arjun Subramanian, Closing the gap: A doctrinal & capability appraisal of the IAF & the PLAAF, Observer Research Foundation, May 02, 2018, https://www.orfonline.org/expert-speak/closing-gap-doctrinal-capability-appraisal-iaf-plaaf/
80. Ibid.
81. Ibid.
82. Saurav Jha, Duel In The Himalayas: How India And China Square Off, *Delhi Defence Review*, July 26, 2017. http://delhidefencereview.com/2017/07/26/duel-in-the-himalayas-how-india-and-china-square-off/
83. Chris Dougherty, Force Development Options for India by 2030, Centre for New American Security, October 23, 2019, https://www.cnas.org/publications/reports/force-development-options-for-india-by-2030

Index